Teaching
Students

with

Special Needs

Using

Google
Workspace
for Education

and

Related Products

By John F. O'Sullivan

Table of Contents

Copyright

Copyright © 2017-2022

by John F. O'Sullivan Jr.

Dedication

I want to dedicate this book to all the teachers trying to integrate technology into their classrooms and helping struggling learners get the best possible education.

Disclaimer

For full disclosure, my brother is an employee of Grammarly. My brother was not consulted for this book.

This book does not accept advertising. I am not associated with the brands that I recommend. None of the content is ads or paid promotion.

My motivation is personal because I was one of the first special needs students in the modern special education system. I was helped by technology. I hope that this book helps educators teach special needs students with technology.

About the Author

John F. O'Sullivan worked as a special education teacher for ten years. He later became a technology integration specialist. He currently works as the Chelmsford High School Librarian and as an Assistive Technology Specialist for Chelmsford Public Schools. He is a Google Level 2 Certified and ISTE certified educator.

John has written seven books on educational technology, including this book. Five of those books are educational technology guides. His books include The UDL Educational Technology Guide 2020, The Educational Technology Guide 2018, The UDL Educational Technology Guide 2018, UDL Technology, The Teacher's Awesome App Guide, and Teaching Projects with Computers.

John was one of the first special needs students in the modern system of special education. As a result, he was four years behind in reading because of educational neglect. John's motivation is out of a need to improve on a system that let him down as a child and with the hope of helping others. He believes the path to success for special needs students includes technology integration, universal design for learning, and creativity.

Twitter: @The_app_guide
email: Jfosullivan71@gmail.com

Introduction

Why write a book about how to use Google Workspace for special education students? Education is taking a sharp turn towards Google Workspace for Education, and there is no sign of looking back. There are many reasons schools have decided to use Google products to teach students. If you have one-to-one devices, Chromebooks are an economical option. Google Workspace for Education also takes feedback from teachers. Some unnamed competitors are not as good at this.

How is the iPad guru authoring a book promoting Google Workspace for Education? I am guilty of writing the most comprehensive iPad guide for educators: The Teacher's Awesome App Guide. I traded my iPhone in for an Android. That book was free, six hundred pages, and downloaded over six hundred times. This book can also have an impact.

I believe that you work with what you have in front of you. With schools committing to Google, teachers need to understand how to use the products in the best way possible. If you want to help educators use Google Workspace for Education for special needs students making a book about Google is a fast way to get techy educators' attention. The goal is to educate teachers on how to help special needs students to succeed with the Google platform.

As a classroom teacher, I understand the power of teaching ways to utilize the right tools. If you want students to use a tool right in front of them, the prospects are much better. Most teachers want to spend time teaching and developing students. They do not become computer experts and salespeople influencing administrators regarding technology procurement. All teachers want to get out more than they put in. The key is to leverage tools to have the most significant impact for a reasonable amount of time.

Chromebook Accessibility

All Chromebooks have great built-in accessibility features. Many people are unaware of these fantastic tools.

How to access these tools:

Click on what I call the oval.

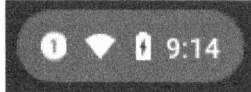

Look for a circle that says accessibility. When you first get your Chromebook, it is defaulted to be off and will look like this. In the top right corner, click on the gear called settings.

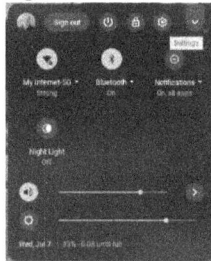

Click on advanced.

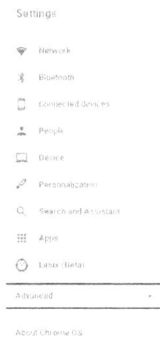

Then Privacy and security.

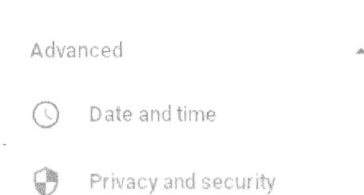

If that is too hard, you can search for accessibility in the search menu.

Next, hit the button to turn on the accessibility menu.

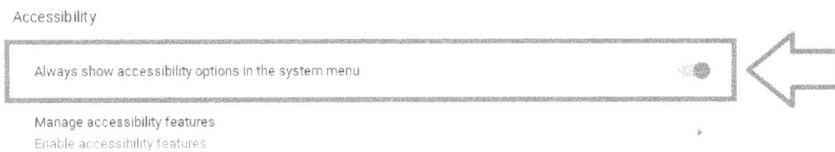

Now the accessibility menu will appear from now on. To get the menu to re-appear, just click in the bottom right corner next to the time. Then click on the accessibility menu.

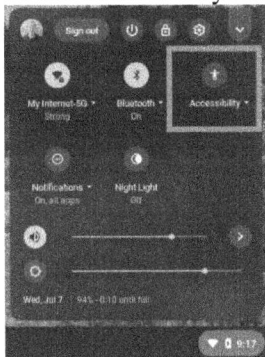

To turn on the assistive technology, click next to the desired item, and a blue check will appear when enabled.

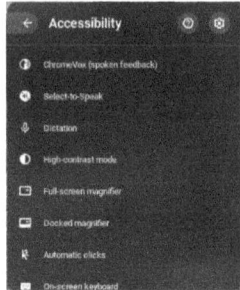

How to Add a Keyboard
Click on the space next to where the clock is located. Then click on the gear on the top right of the pop-up window.

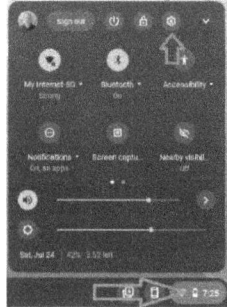

Click on the device.

Click on the keyboard.

Scroll down and click on "change input setting."

Click on "add input methods."

Search for the language, then check the box. Finally, click add.

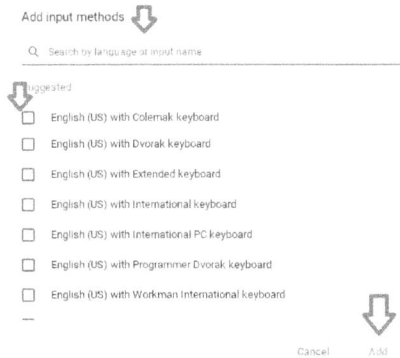

Add input methods

Q   Search by language or input name

Suggested

☐   English (US) with Colemak keyboard

☐   English (US) with Dvorak keyboard

☐   English (US) with Extended keyboard

☐   English (US) with International keyboard

☐   English (US) with International PC keyboard

☐   English (US) with Programmer Dvorak keyboard

☐   English (US) with Workman International keyboard

Cancel   Add

If you need a short cut type the words "language inputs" into the search menu in your Chromebook settings.

Click on the space next to where the clock is located. Then click on the gear on the top right of the pop-up window.

Type "language Inputs"

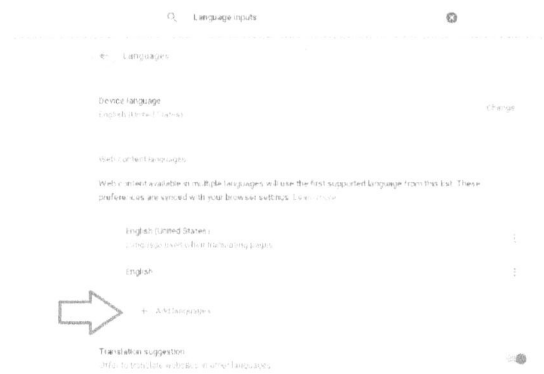

Then click on add language and search for or pick the appropriate language.

Chapter 1 Google for Reading and Writing

If you can see text on a computer screen, it can be read to you for free from a program that comes with most operating systems. The programs that read text are called speech-to-text. In Chrome, there are two programs Select to Speak or ChromeVox that read text. For Windows, the program is called Navigator. The Edge browser has Read aloud an amazingly effective program. For Apple users, it is called Speak Selection. In the past, you had to buy a particular program or service to have text read to students with disabilities. Now text can be read on the device for free, and it is integrated with the computer. No clunky device, and since integrated into the operating system, the feeling of being different is diminished. If you are accessing Google products from the Chrome browser, you can have the text read with the help of free text-to-speech extensions as well.

All the major operating systems have free speech-to-text programs that integrate with the computer. For Chromebooks and Apple products, it is called Dictation. In Windows 10, it is called Dictate. You need a microphone to use the programs. However, most computers come with free built-in microphones. If you somehow buy a computer that does not have a microphone, you can buy an effective microphone for a very reasonable price.

There is no reason why a student that wants or needs text-to-speech or speech-to-text cannot get it for free on the computer that they have and use. Indeed, there are alternatives that you can download to give you more options. All you must do is enable the feature in the computer's settings in most cases.

Chromebooks

Select to Speak

Free on all Chromebooks is a text-to-speech program called Select to Speak. It is not an extension, and you do not have to download the program. You must go into the setting to enable the program. See the directions below to enable this feature.

Dictation

Free on all Chromebooks is a speech-to-text program called Dictation. You just must go into your settings and enable it. What is great about this program is that you can use speech to text on many websites. See the directions below. All Chromebooks have great built-in accessibility features.

How to access these tools:

Click on what I call the oval.

Look for a circle that says accessibility. When you first get your Chromebook, it is defaulted to be off and will look like this. In the top right corner, click on the gear icon called settings. (top right)

Click on advanced.

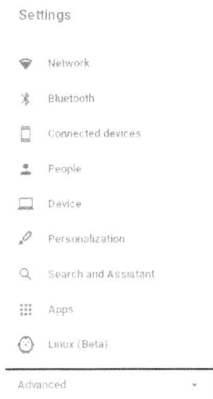

Settings

🛜 Network

⁑ Bluetooth

▢ Connected devices

👤 People

▭ Device

✎ Personalization

🔍 Search and Assistant

⁝⁝⁝ Apps

⬡ Linux (Beta)

Advanced   ▾

About Chrome OS

Then Privacy and security.

Advanced   ▴

🕐 Date and time

🛡 Privacy and security

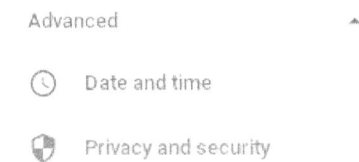

If that is too hard, you can search for accessibility in the search menu.

🔍 Accessibility   ⊗

🧍 **Accessibility**   →

🧍 Manage **accessibility** features   →

🧍 Always show **accessibility** options   →

Next, hit the button to turn on the accessibility menu.

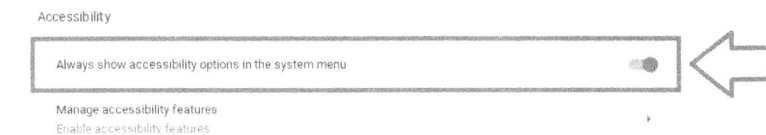

Accessibility

Always show accessibility options in the system menu   ⬤   ⇦

Manage accessibility features
Enable accessibility features   ▸

Now the accessibility menu will appear from now on. To get the menu to re-appear, click in the bottom right corner next to the time. Then click on the accessibility menu.

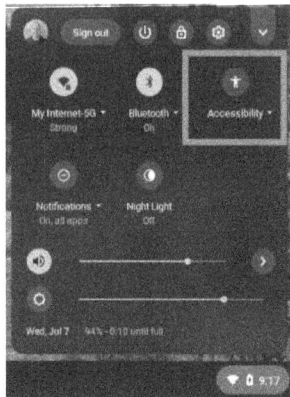

To turn on the various assistive technology, click next to the desired item, and a blue check will appear when it is enabled.

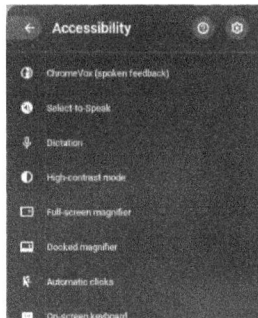

To turn on ChromeVox, click on the line where it says ChromeVox. This is a text-to-speech reader. The downfall to ChromeVox is that it reads everything on the screen. Most people want a particular passage read to them. The majority of people will prefer Select to Speak.

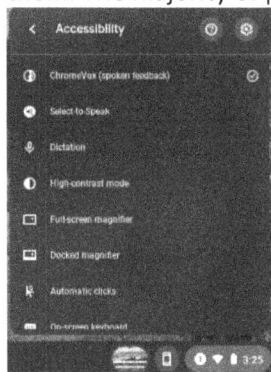

Select to Speak is a text-to-speech feature on all Chromebooks. Just click on where it says Select to Speak to enable the program and get the blue check mark.

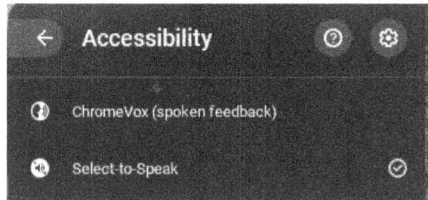

First, click on the icon in the bottom right corner.

Click and drag over the text that you want to have read to you.

This is the most comprehensive educational technology guide for special education. The best technologies are highlighted in each chapter. You can completely transform your teaching practices with the technologies within this book. You will learn about the best technologies for executive function, visual impairment, learning disabilities, speech and language, and technology for engaging hands-on projects for special needs students. Only the very best technology makes the book. The information in the book is based on many years of research.

The program will highlight what it is reading.

This is the most comprehensive educational technology guide for

When using Google Documents, there is a free feature called Voice Typing. (speech-to-text) Under tools, scroll down to Voice Typing.

Click on the microphone icon and talk. When you are done talking, click on the microphone icon.

When searching with Google, there is a built-in microphone. Just click on the microphone and talk.

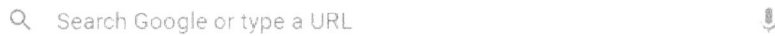

To enable Dictation, click on the line where it says Dictation.

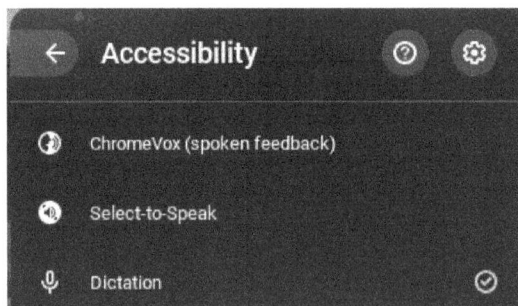

Then click on the microphone icon next to the clock.

Dictation works with website-based products. When you are on a Chromebook, Google Workspace for Education is web-based, and other online programs that you can use as well.

What if you have a PC?

Dictation on Windows

Dictation is a free program that comes with Windows 10.

Directions:

Windows Key + H

To use dictation, go to Settings and turn on online speech recognition

You click on the link and change the settings.

You must enable online speech recognition the first time. Type the word Dictate in the search menu.

Click on the button to enable the feature.

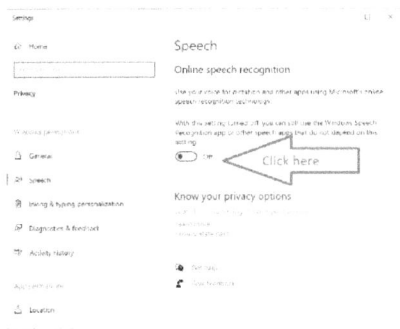

You might have to adjust the microphone settings. Go to the search menu and type the word microphone, then click enable.

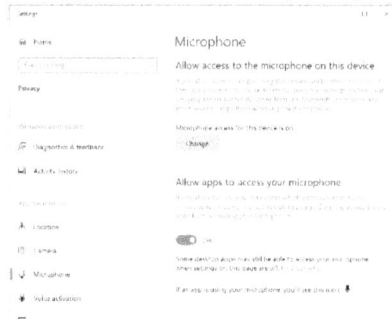

To find settings click on the Windows icon in the bottom left corner.

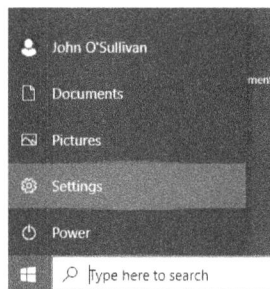

The free text-to-speech program on Windows is called Narrator. The program works fine but tends to read everything. If you have an entire book read in the browser to you, then I recommend it. If you want a sentence or two read to you, then it might not be worth it.

Go under settings in Windows. You can also search in the search menu for Narrator.

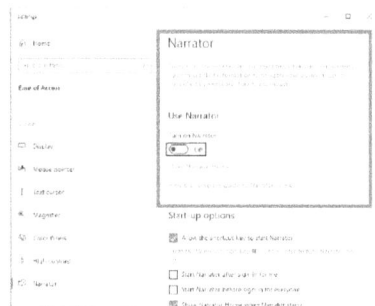

There is a free called ReadAloud or Read Aloud by Microsoft. Not to be confused with a Google extension of the same name. You must go to the Windows Store to download it. The program works great. However, it is designed to work with Microsoft products. If you use the Edge browser, this is an outstanding program.

Google Extensions

Google does not have a perfect reading and writing program to fix all your problems. However, the good news is that Google is such a robust platform that many products integrate with Google.

Extensions – These are simple Chrome browser plug-ins that perform many valuable functions.

The first group that integrates with Google is extensions. These are simple products that you can install in the Chrome browsers from the Google Web Store. Schools that use Google can block some or all extensions. If your school does not block extensions, this will be an effortless process. If the extensions are blocked, you will have to communicate with your school's information technology department. The good news is that most schools want teachers to experiment and use technology. Knowing they have a personal stake in your success should encourage you to communicate with the technology department.

Chrome Web Store

https://chrome.google.com/webstore/category/extensions?hl=en

Text to speech – These are programs that read the text to you.
Text to speech is on every platform, is typically free, and is quite easy to use. This type of software is on all operating systems and browsers. This would include Google products. All levels of computer users can use this type of program.

Look at the graphic below to see what to do.

Chrome

Readme

This program reads different file types, including PDFs. However, this extension will read many different types of documents. You must upload the documents that you want to read.

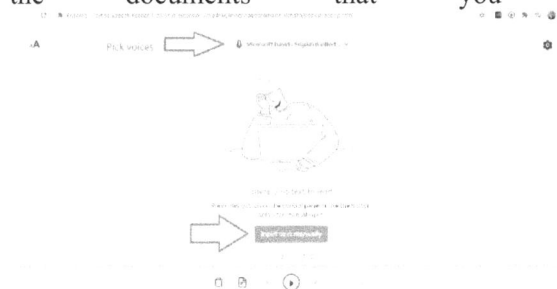

https://chrome.google.com/webstore/detail/readme-text-to-speech-rea/npdkkcjlmhcnnaoobfdjndibfkkhhdfn

Read Aloud: A Text to Speech Voice Reader

This extension has a very human-sounding voice.
https://chrome.google.com/webstore/detail/read-aloud-a-text-to-spee/hdhinadidafjejdhmfkjgnolgimiaplp

Select to Speak - Text to Speech

To use Select to Speak, you must click on the extension that looks like a play button on the top right of the browser. The extension reads Google Documents. Select to Speak can also be used to read web pages.
https://chrome.google.com/webstore/detail/select-and-speak-text-to/gfjopfpjmkcfgjpogepmdjmcnihfpokn

Natural Reader
This is a text-to-speech extension.
https://chrome.google.com/webstore/detail/natural-reader-text-to-sp/kohfgcgbkjodfcfkcackpagifgbcmimk?hl=en-US

ChromeVox
This extension is preloaded on all Chrome devices. ChromeVox includes Chromebooks and Chrome boxes.
The extension can be added to the Chrome browser on a PC or Mac computer.

The extension acts more like a professional assistive technology as opposed to just a basic text-to-speech extension. The extension is designed for people with visual impairments. ChromeVox will read the entire screen to you. This is helpful if you cannot see the screen. However, if you just want selected words to read to you, this can be a distraction. The extension is valuable if you understand the population that it is intended to be used.

http://www.chromevox.com/

Click on what I call the oval. (ChromeVox on a Chromebook)

Click on the accessibility menu.

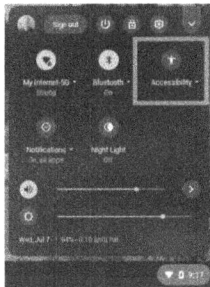

Click on ChromeVox, and you will get a blue checkmark.

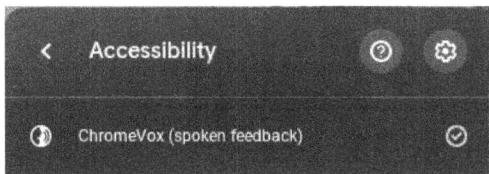

Snap&Read
This app simplifies the webpage that you are reading, has optical character recognition, and reads to you. Snap&Read is a popular accessibility extension.
https://chrome.google.com/webstore/detail/snapread/mloajfnmjckfjbeeofed aecbelnblden

Mobile Phone

Both Google and Apple have accessibility features already installed. If you go into the setting and then accessibility, you will find text-to-speech, speech-to-text, and features for the visually impaired, including magnification. On Apple devices, text-to-speech is called Select and Speak. The same is true for iPads and Apple laptops. You will find many of the assistive technology features that you need on Apple products. You can search the App Store for more assistive technology apps. I recommend using Google to search for the apps instead of using the search feature in the App Store.

On Android phones, you can go to the settings and search accessibility. You will find, in most cases text-to-speech, speech-to-text, and settings that can be changed for the visually impaired, including magnification. Android phones have several apps that you can download and install. The Google Play Store is well organized and has good reviews. I would suggest that you search the store and read the reviews and summaries of what each app does.

Google OCR (Optical Character Recognition)

PDFs can be converted to text with Google OCR. The option is free and works very well. Once the PDF is converted to text via Google OCR, a text-to-speech reader can read it. Google OCR is very accurate. However, OCR readers can make occasional mistakes. If you are using a lousy photocopy that is scanned in, expect more mistakes.

Optical character recognition or OCR can convert inaccessible text to accessible text. This could mean several different things. Google OCR converts PDFs, JPEGs, and handwritten notes to text. You would still need another product to read the text.

Alt plus left click on the touchpad. Click on "open with." Then open as a Google Document. (Note works with images and PDFs)

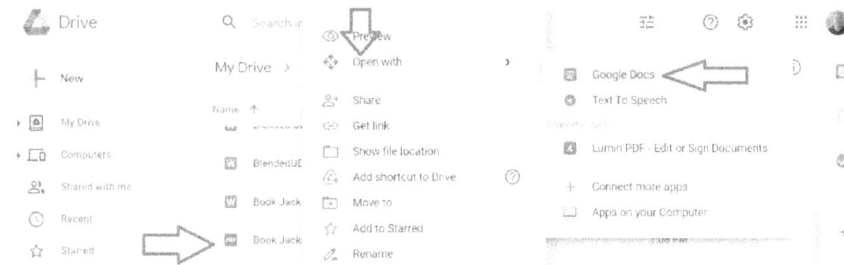

Note: Google OCR works with pictures as well. If you take a photo or screenshot with text, follow the directions above. Google OCR will convert the image to text. Google OCR will convert handwriting and pictures of text to accessible text.

Google OCR Cloud Version

https://cloud.google.com/vision/docs/ocr

Google OCR Support

https://support.google.com/drive/answer/176692?co=GENIE.Platform%3D Desktop&hl=en

Reading Services for the Disabled

Possibly the most underutilized services in special education are book reading services for the blind and dyslexic. Bookshare is free for individuals with qualifying disabilities. I used Learning Ally when I was in college in the early 1990s. At the time, it was called Recording for the Blind and Dyslexic. The use of books on tape for me was life changing. However, Learning Ally does have an annual fee. The Learning Ally's books are eBooks that are read to you today.

The biggest argument against using these services is students with disabilities do not want to be different. When I was in college, I decided I was going to use these services. I felt a need to speak about it. I remember telling my roommate and some of my friends about it. No one cared. When people are less afraid of failure than being different, something is wrong. The truth is that people with disabilities do not want technology because of how they feel. For the most part, people do not care. However, people with disabilities often do care about how they are perceived.

The technology is seamless today. You can use these services in a browser or an app. Both Learning Ally and Bookshare essentially look like eBooks. Some of the older books for Learning Ally are just audio. Mostly, someone observing a student using one of these services would see an eBook. Of course, the student could use headphones to hear the audio. If they feel uncomfortable, they can tell others they are listening to music. If you access the book on your phone with earbuds, it looks like you are listening to music.

Bookshare

Bookshare is a free service for those that qualify. Bookshare works with the Chrome browser and other mainstream browsers. Keep in mind not all disabilities qualify for the service. Dyslexics and the visually impaired typically qualify. Autism and intellectual disabilities do not qualify. I would suggest reading Bookshare's website or calling them with questions. When I have called Bookshare they have been very helpful and kind.

https://www.bookshare.org/cms/

If you are using a phone or tablet, several apps work with Bookshare. The service can be used on any device. However, some apps and extensions will add functionality.

Google Extensions

Capti Voice

This is a text-to-speech extension that works with Bookshare.

https://chrome.google.com/webstore/detail/capti-voice/nlngjmhlfdekmgoaaiendhkcbhdcedjh

BeeLine Reader

This extension makes reading the words easier. This is excellent for people with reading issues.

https://chrome.google.com/webstore/detail/beeline-reader/ifjafammaookpiajfbedmacfldaiamgg

Google Play

Go Read

This is a free app that works with Bookshare.

https://play.google.com/store/apps/details?id=org.benetech.android

Dolphin EasyReader

This is a good free option if you don't have an app budget.https://play.google.com/store/apps/details?id=com.yourdolphin.easyreader

Legere Reader

This is a paid app that gives you added functionality. If you use the service on a regular basis, this app is worth consideration.

https://play.google.com/store/apps/details?id=voicedream.reader&hl=en_US

iPad or iPhone

iPhone or iPad (Apple Products)

Voice Dream Reader

This is the app that I have recommended for years.

https://www.voicedream.com/reader/

Dolphin EasyReader

This app is free. If you do not have an app budget, this is a good option.

https://apps.apple.com/us/app/dolphin-easyreader/id1161662515

Learning Ally

You can read books from Learning Ally with the Chrome browser or other mainstream browsers. Dyslexics and the visually impaired typically qualify. Autism and intellectual disabilities do not qualify. I would suggest reading the Learning Ally website or calling them with questions. Learning Ally is also a paid service. The annual fee varies. If you are looking for the current cost, I recommend looking at their website or contacting them directly.

https://learningally.org/

OverDrive

Most of the public libraries have OverDrive. Library patrons can take out library eBooks and audiobooks online. Some public schools also have this service. Many times, it is hard to maintain engagement with text-to-speech readers. Audiobooks recorded with a professional reader can increase engagement.

https://www.overdrive.com/

Sora

This app allows students to download an eBook or audiobook onto their phones. The app is for both iPhone and Android phones.

https://www.overdrive.com/apps/sora/

Teaching Reading

Epic
This is a great website with a ton of books for elementary students. Best of all, the resources are free for teachers.
Website
https://www.getepic.com/

Newsela
Newsela allows you to adjust the reading level to the student. You can have several students read the same article on different reading levels.
https://chrome.google.com/webstore/detail/newsela/bfpeiapdhnegnfcfkdfihabadngjagfj?utm_source=chrome-ntp-icon

Library Extensions
This extension helps you access your local library as you visit different websites. This is an excellent way to save money for teachers and students that are looking for some books.
https://chrome.google.com/webstore/detail/library-extension/chkgcmmjoejpekoegkedcpifgfhpjmec?hl=en

Teaching with Read & Write

Read & Write is an often misunderstood program. People sometimes buy the program because they want a text-to-speech and/or a speech to text program. Chromebooks come with exceptionally good speech-to-text and text-to-speech already built-in. All platforms have free assistive technology options. Both Windows and Apple have free options for both. Please do not get me wrong Read & Write is an excellent program that I highly recommend. If you want a word prediction program Read & Write is an excellent choice. If you want a powerful assistive technology program that does many remarkable things, Read & Write is for you.

This is a YouTube video tutorial on how to use Read and Write that I created.
https://www.youtube.com/watch?v=xtBbbmaKLTo

The biggest secret to Read & Write is that it works with several other extensions. Everyone knows about the Read & Write extension. There are others that you can download and use as well. You should also download OrbitNote, formally known as Texthelp PDF Reader. The other extensions associated with Read & Write from Texthelp are listed further down in the reading.

OrbitNote

Note: Texthelp PDF Reader changed its name to OrbitNote. This was a good marketing decision. The extension is more than a PDF Reader. OrbitNote formally known as Texthelp PDF Reader does what it says. The program will read PDFs. Just remember that no PDF reader can read all PDFs. The best way to explain it is that if you turn a Google Document or a Word Document into a PDF, this program will read it. However, if you photocopy or scan a PDF, most PDF programs will struggle to read such a document, including this one. The more pictures, text boxes, and formatting, the more difficult it will be for any program to read. However, this extension is still excellent. Texthelp has several great tools that allow you to make changes to PDFs to be more accessible and readable. The OrbitNote extension should be used to teach, create lessons as well as make PDFs accessible. That is probably the reason for the name change. OrbitNote is a teaching resource with many outstanding assistive technology features and one of those tools is a PDF reader. I recommend that everyone that uses Read & Write and OrbitNote to instruct your students regularly.

The point of the Texthelp extensions is to teach all students. The idea is that all the students should have the extension, and the teacher(s) should use the technology. OrbitNote allows both the teacher and student to annotate a PDF. If the teacher creates annotations on a document, the student(s) can see the notations if they have the same extension and license. OrbitNote allows the teacher and student to write on the PDFs, Type, and record your voice. You can use the speech-to-text tool, word prediction, highlighting, and spell check when you type on the document. Put all these tools together, and you have something impressive. Imagine recording directions for students that have reading or attentional problems. Annotate what the students are going to read to point out the essential information. Then the students can use word prediction and speech-to-text to communicate their answers. That is breaking down barriers.

Tutorial OrbitNote
This is a tutorial that I created.
https://www.youtube.com/watch?v=woB6cpMM588

Texthelp YouTube Channel
Texthelp has a great YouTube channel.
https://www.youtube.com/channel/UC50ySjJFHy9ZJltP6S0AlHg

Mote
Mote is an alternative to OrbitNote. Mote gives students and teachers the ability to leave recorded comments in Google Workspace. With Mote, you can create comments in Google Documents, Slides, or other programs in Google Workspace that can be recorded and shared. This is an outstanding extension. There is a free and paid version.
https://chrome.google.com/webstore/detail/mote-voice-notes-feedback/ajphlblkfpppdpkgokiejbjfohfohhmk?hl=en-US

Read & Write

Read & Write (Chrome Extension)

Read & Write is an excellent word prediction program. The primary advantage is that you can use it across settings. Logging into Chrome to access the program circumvents the need to load the program on multiple computers. With one license, you can use the program at home, school, and other settings. Best of all you can access the program on a computer that you have never used before.

Read & Write has a large number of functions that do just about everything you need. The program is a Swiss army knife of accessibility options. Read & Write has a speech to text, text-to-speech, dictionary, picture dictionary, and highlighting/note-taking abilities, just to name a few.

A teacher can get this extension for free. This allows teachers to test out the product.
https://www.texthelp.com/en-us/products/read-write/free-for-teachers/

Install the following extensions
Go to the Chrome Web Store and load the following extensions:

Read & Write
https://chrome.google.com/webstore/detail/readwrite-for-google-chro/inoeonmfapjbbkmdafoankkfajkcphgd?hl=en-US

OrbitNote (Texthelp PDF Reader)
https://chrome.google.com/webstore/detail/texthelp-pdf-reader/feepmdlmhplaojabeoecaobfmibooaid?hl=en-US

Screenshot Reader
This extension reads selected content on a webpage. This extension can be used with Read & Write.
https://chrome.google.com/webstore/detail/screenshot-reader/enfolipbjmnmleonhhebhalojdpcpdoo?hl=en-US

EquatIO
This is a Google extension by Texthelp for math. This is unrelated to Read & Write. I recommend trying this extension.
https://chrome.google.com/webstore/detail/equatio-math-made-digital/hjngolefdpdnooamgdldlkjgmdcmcjnc?hl=en-US

Read through the list of Read & Write extensions. The link also tells you how to install the extension and basic troubleshooting. If you scroll down, they have a good guide about using the Read & Write tools.
Read & Write Tools
https://www.texthelp.com/resources/training/read-write/google-chrome/

Start with the first tool on your far left, which is called the spell-checker.

Type a sentence and make a few mistakes. Then click on the checkmark to make a few corrections.

Use the second tool on the left called the Prediction

Type a sentence. Use the tool to spell words. Click on the suggested words to hear them.

Use the Dictionary tools.

Highlight several words and get a definition. Make sure the dictionary reads a definition to you. Click on the picture dictionary as well.

Find an online article to test this tool. You can also use this article.

Police official: Short-circuit likely caused Notre Dame fire

https://www.apnews.com/c7e795958c894ef491201aa9ef835638

Use the speech-to-text tools.

Highlight something you want to have read to you.

Use the Screenshot Reader tool.

Highlight a paragraph in the story and have the tool read to you.

Read the same article with the Screen Mask.

Click on the Speech Input tool. Then say a complete sentence to your Chromebook.

Highlight a word and translate that word into a new language with the Translator tool.

Highlight several words in different colors with the highlighting tools.

Now that you have highlighted several words, click on the Collect

Highlight tool.

Next, click on the vocabulary tool.

Clear all the highlights with the Clear Highlight tool.

Leave a voice note with the Voice note tool in a Google Document.

Practice reading out loud with the Practice Reading Out Loud Tool.

These are extensions that are commonly used with Read & Write.

Snapverter

This web service works with Read & Write. Snapverter integrates with Google Drive. You upload a document to the Snapverter folder in Google Drive, and it converts the document to a format where the text can be read by Read & Write.

https://www.texthelp.com/en-us/products/snapverter/

OrbitNote (Texthelp PDF Reader)

This app will read and annotate PDFs. The app is often used with Read & Write from Texthelp. This is an extension that is used with Read & Write to edit PDFs.

https://chrome.google.com/webstore/detail/texthelp-pdf-reader/feepmdlmhplaojabeoecaobfmibooaid?hl=en-GB

Open this PDF with OrbitNote. When you open the PDF at the top, it will ask you to open it with Read & Write. You also can click on OrbitNote and open a PDF from Google Drive.

Go to the icon on your far left of the toolbar.

Highlight a sentence or a phrase and click on the icon.

Next, go to the dictionary tools.

Highlight a word and use the dictionary. Make sure that it reads a definition to you. Then click on the picture dictionary.

Use the text-to-speech tool.

Highlight a paragraph and have it read to you. Practice pausing and stopping the passage.

Next, use Screenshot Reader.

Click on the icon and highlight a paragraph.

Highlight a word and click on the translator tool.

Now highlight several words and phrases with the Highlighting tool.

Now click on the Vocabulary tool.

Click on the Typewriter tool.

Type a note to yourself next to a sentence that you find interesting.

Use the Push-Pin tool.

Click on the tool and then click next to where you want to make a note. Then click outside the box to close.

Next, use the Freehand Drawing tool.

Circle a sentence in the passage. Next, change the color and write your name.

Use the Shapes Drawing tool to draw a red square and a blue circle.

Chapter 2 Google for Executive Function

What is executive function?

Executive functioning is defined in many ways. The definition is vast and abstract. Executive functioning is about organizing, prioritizing, remembering, and emotional regulation. Being able to transition between tasks is also a function of executive functioning. This encompasses many cognitive abilities that develop with age.

I work with teenagers. Some of the behaviors of teenagers can be explained by developing executive functioning. The part of the brain that regulates behavior has not fully grown yet. The part of the brain that regulates behavior is not fully grown until about 25 years of age. For people with a delay, this can take longer. Of course, some adults cannot fully control their behavior. Executive functioning is developmental, and it cannot be forced to happen sooner. Incremental gains can be achieved with structured lessons and reinforced behavior. However, an underdeveloped brain that is growing cannot be solved with excellent teaching or parenting.

How common are executive functioning deficits? If you look at the number of diagnosed people with ADHD or ADD, the numbers vary by year. Then the criterion for diagnosis is not exact. One doctor might give you medication because you think you have it, and another needs to see severe issues before treatment. A good measure is looking at students' book bags or ability to complete tasks. If you are looking at special needs students with this standard, then the number is high. If you are looking for people treated with ADHD or ADD, then the number is small. However, using technology to help students with executive functioning issues make your life easier and helps your students. As educators, if we can improve how our students perform an everyday tasks that is valuable. Many students have a weakness in some areas of what we call executive functioning. Using technology as an aid will improve their achievement in school.

Why should You Put Documents Online?

There are many reasons why teachers should post written directions online. What you will find when you look are many students with some sort of issue that makes written directions necessary. Some of the difficulties students might have include speech and language deficits, anxiety, and executive functioning issues. Executive function deficits can consist of memory issues, self-awareness, inattention, organization, ability to prioritize, and self-regulation. What makes this problem even worse is the fact that students often have multiple issues that combine to make the need for written directions and having all assignments posted online important.

Executive Function

Executive function is a cornerstone of blended learning and universal design for learning. Let me explain. First, you must understand what executive function is? Most definitions call it the ability to organize and prioritize. This is a very abstract concept that can be many things. For the sake of school, we are talking about writing down your homework, remembering due dates, knowing what to study, prioritizing, understanding, time management, remembering, organization, attention, and many other tasks. It should be no surprise that students who have strong executive functions do very well in school, and those without the skill lag behind.

If an executive function is a strong prediction of school success, why is this not at the center of the field? It is thought that students will learn this independently, and some people believe that we can teach these skills. The truth is that these skills are challenging to teach. The reason is that these skills are developmental. You cannot give the exact directions to a first grader that you would give to an eighth grader. Some people have a much stronger ability in this area, and some do not.

Often what we are talking about when we talk about executive function skills are language skills. If you have difficulty with receptive language, then you will miss essential directions that you need. We tend to assume that everyone can process the information when we give directions and then know what to do with this information. Then we assume that students will ask questions. There are many students that lack expressive language.

The next issue that has to be considered is self-awareness. Often students that have bookbags that are a mess and do not write down their homework do not see the extent of the problem. In some cases, they do not even see a problem. If you are not aware that you have a problem, you are not able to fix it.

Memory issues compound this. You have to remember any number of things during a school day. If you forget your textbook and do not do your homework, this can make your issues even worse.

The hardest part of all of this is that many students have multiple problems. All of this combines for lower grades with no end in sight. Dealing with students that do not follow directions, bring materials, and do their homework can take a good deal of time and energy of a teacher that is trying to help.

With all the frustration that executive function and related problems causes, it is shocking that everyone is not using technology to address this problem. All your papers, assignments, due dates, and materials should be online in Google Classroom. All my work is online with the directions and material that students need. Students, parents, special educators, English as second language teachers, and counselors can see the materials if the student logs into the account. Classroom teachers can add any teachers they want to the account.

Most adults add information to the electronic calendars, and we should expect the same of students. If you use paper, you are a shrinking minority. There was a time when all we could do is put information on paper, and there were no other options. Students had to learn how to fill out a planner, carry a giant bookbag, and have no other choice. As a result, we told students that they had to learn how to do this. Those days are long gone. In business, file sharing is the rule, and printing is the exception.

One of the most significant barriers to learning is executive function difficulties often, compounded with a lack of self-awareness. Using a program like Google Drive to share documents, and Google Classroom to organize papers and assignments is the best thing to happen to executive function ever. Uploading a document to Google Drive or sharing a document in Google Workspace is about as complex and costly as an email attachment. The fact that the papers with basic directions are online is a game-changer. Many students are too shy or have anxiety about asking a teacher a question. If the directions and material are online, the student can read the document and figure out the assignment. Then if they have a question, they can ask about the document. Reluctant students can show the document to adults that they feel will not judge them like parents or special educators. The amount of stress that simple uploading documents can alleviate is tremendous.

Sharing documents online is not time-consuming, complicated, or expensive. Offline documents are not accessible. Once documents are put online, they can be better accessed by students with any number of disabilities. Executive function difficulties are a fairly common problem that is often ignored.

References Executive Functioning

Arain, M., Haque, M., Johal, L., Mathur, P., Nel, W., Rais, A., Sandhu, R., & Sharma, S. (2013, April 3). *Maturation of the adolescent brain*. Neuropsychiatric disease and treatment. Retrieved August 4, 2022, from https://www.ncbi.nlm.nih.gov/pmc/articles/PMC3621648/#:~:text=There%20are%20several%20executive%20functions,development%20of%20the%20prefrontal%20cortex

Belsky, G. (2022, July 18). *What is executive function?* Understood. Retrieved July 31, 2022, from https://www.understood.org/en/articles/what-is-executive-function

Center for the Developing Child Harvard University. (2020, March 24). *Executive Function & Self-regulation*. Center on the Developing Child at Harvard University. Retrieved July 31, 2022, from https://developingchild.harvard.edu/science/key-concepts/executive-function/

Psychology Today. (n.d.). *Executive function*. Psychology Today. Retrieved July 31, 2022, from https://www.psychologytoday.com/us/basics/executive-function.

You only need these skills to use a planner!

1.   **Receptive language** to listen to the teacher when it is time to write down homework.

2.   **Expressive language** so that you can ask a question when you do not understand the homework.

3.   **Self-awareness** that you need to use a planner and write down work in a planner.

4.   Students must **remember** to read the planner, bring planner to school, to home, bring a pencil to write in a planner, and do what it says in the planner.

5.   **Fine motor coordination** to write it down so that you can read what you wrote.

6.   **Abstract reasoning** to understand what is essential and needs to be written down

7.   **Ability to focus** when the teacher is talking

8.   Able to **see** the board and able to read teachers' handwriting

9.   **Motivation** to do all the above after one or more of these areas fails

10.   Some students have **anxiety**. Many students are afraid to ask a question. Public speaking is the number one fear that people experience. (Ladouceur 2022) Even students without anxiety can be afraid to ask.

Ladouceur, P. (n.d.). What we fear more than death. Mental Help What We Fear More Than Death Comments. Retrieved July 16, 2022, from https://www.mentalhelp.net/blogs/what-we-fear-more-than-death/

Do your students have problems in any of these areas?

1.   Receptive language

2.   Expressive language

3.   Memory

4.   Hearing

5.   Seeing

6.   Fine motor coordination

7.   Self-awareness

8.   Organization

9.   Motivation

10.   Anxiety

Anxiety Disorders

The prevalence of anxiety disorders is shockingly high. According to the National Institute of Mental Health, 19.1 percent of people in the United States of America have had some form of anxiety disorder in the past year. We are talking about ever having an anxiety disorder; that number jumps to 31.1 percent. The prevalence of anxiety disorder is more common in women, with a rate of 23.4 percent. For men, the rate is 14.3 percent. (National Institute of Mental Health 2019)

What does this mean for educators? Teachers should be concerned about a student having an anxiety attack. Often teachers see none of the signs. A child might go home and express anxiety at school and seem fine in front of a teacher. Educators need to understand that anxiety alters behavior. Children with anxiety might be unable to ask questions out of fear. This makes clear directions even more critical. This is why posting all of your homework and assignments online is paramount. If an adult can easily get a copy of your directions or assignments, they can read them or explain the information to the student. Educators cannot fix anxiety disorders. We can show that we are doing our part to help students.

Furthermore, a student who does not know what to do but is scared to ask can read the directions and figure out the assignment. If a student does not understand what to do for the homework, they now have a reference point to ask for help with online written directions.

Language Disorders

How common are speech and language disorders? "Nearly 1 in 12 (7.7 percent) U.S. children ages 3-17 has had a disorder related to voice, speech, language, or swallowing in the past 12 months" (The National Institute on Deafness and Other Communication Disorders 2017) The number does not sound that large on the surface. Students missing important information because of receptive language difficulties is quite common. Having highly effective communication is a valuable skill for everyone. As teachers, we do not want to think that we are not effectively communicating with students. Many teachers give responsibility for listening to the students. It is frustrating to repeat yourself. Students in regular education classrooms are trying to listen yet missing important information. That is why I recommend posting the information online where parents, special education teachers, English as a second language teachers, and students can access it.

National Institute of Mental Health. (n.d.). Any Anxiety Disorder. Retrieved December 30, 2019, from https://www.nimh.nih.gov/health/statistics/any-anxiety-disorder.shtml

The National Institute on Deafness and Other Communication Disorders (NIDCD), part of the National Institutes of Health (NIH). (2017, December 21). Quick Statistics About Voice, Speech, Language. Retrieved December 30, 2019, from https://www.nidcd.nih.gov/health/statistics/quick-statistics-voice-speech-language

Google Classroom

How does Google Classroom help students with executive functioning?

Google Classroom organizes and prioritizes the work. It is considerably harder to lose your work with Google Classroom. Students can post comments, and the teacher can answer questions. Students can ask questions privately. Teachers can document that they answered questions or gave a reminder of work to students.

If you have ever seen a social media post and thought, what were they thinking? Then you can understand this principle. There are many barriers to communication online. You cannot see or hear the people in the conversation. As a result, people are less afraid of saying the wrong thing. This principle also works for people with anxiety that are afraid to ask a question for fear of being judged. Being less inhibited when they must ask a question is a positive. Public speaking ranks above death as a fear. This is not true when texting or posting on social media. (Ladouceur 2022)

Ladouceur, P. (n.d.). What we fear more than death. Mental Help What We Fear More Than Death Comments. Retrieved July 16, 2022, from https://www.mentalhelp.net/blogs/what-we-fear-more-than-death/.

Some students feel comfortable typing a question in Google Classroom where no one will see or judge them. A teacher can train students to add comments to work. Students can even address the comment towards teachers by typing their email with the @ or + symbol before the email. People feel uninhibited when they post online because they don't hear someone's affect (voice) or see their face. This can be an advantage for a student with anxiety by reducing the stress or fear in communication.

When students hand in assignments in Google Classroom, and we grade and return the work, there is an opportunity to comment. My rule is that I say something positive or give an opportunity for improvement. When a student does not do well on an assignment, I give them feedback.

Sometimes I have them redo the assignment. I find that many students want to redo the work. Students are often thankful that I am giving them more work. I see this an giving a student an opportunity. Some teachers strongly disagree with idea. I had success with the having students on occasion redoing their work.

With Google Classroom, you can do many very positive things concerning executive function. Every time you create an assignment, the date shows up on a calendar within Google Classroom. Google Calendar is designed to allow you to look at many calendars, all or just one. This beats the old days of writing everything down. Google Calendar and Google Classroom both have apps for smartphones. Google Calendar helps students keep track of important dates. Google Classroom stores all the documents and assignments for students in one location.

You can even ask the teacher questions. Google Classroom has a function to share information with parents called Google Guardian. This access to students' documents and assignments is beneficial for parents trying to keep students accountable. Teachers that work in the district can also get access to Google Classroom. This is helpful for special education teachers. With file sharing, you never have to experience that moment again when the printer breaks down, and anxiety increases. Offline and paper documents are not easily accessible. As educators, we want to include everyone and meet their needs.

Google Keep

This is a simple note-taking program. You can set reminders for yourself. The program allows you to color-code your notes. The best part of Google Keep is the fact that you can import your notes into Google Documents. This will enable you to use Google Keep as a graphic organizer. Students can type their notes once and then reshape their ideas in the first draft of their writing.

Open Google Keep here ⇨

To add text, drag and drop or click on the three dots and then click on add to document.

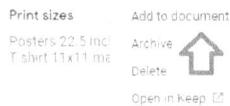

Click on the dots to add text ⇩

Drag and drop comments ⇨

Add to document

Print sizes | Add to document
Posters 22.5 incl | Archive ⇧
T shirt 11x11 ma | Delete
| Open in keep

https://keep.google.com/

Google Calendar

A student can keep multiple calendars on Google Calendar. Each class, a student, takes has a calendar associated with it. When a teacher creates an assignment, it goes onto the calendar. The Google Calendar app is easy to download onto a phone or tablet. This can make keeping track of dates easy.

Click on the box next to the calendar(s) you want to display.

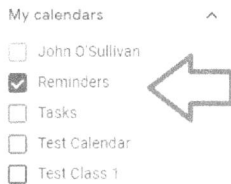

Note: You can display as many or few calendars as you desire. All Google Classrooms create a calendar for each class. When you add assignments with due dates, it appears on the calendar.

To show just one calendar click here.

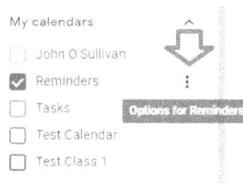

Click display this only

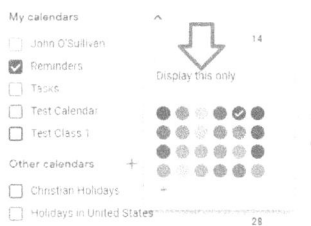

Note: Keep in mind each calendar is color-coded. Each calendar will have text written in a different color. You can also change the default color.

https://www.google.com/calendar

Google Drive

The one thing that Google does well is file sharing via Google Drive. If you ever had a printing issue, you know the stress it causes. It is tough to lose electronic files. You can organize all your files into folders. Google Drive gives you the ability to share your documents for collaboration and grade purposes. If a school keeps all its documents on Google Drive, this can save money. Having the hardware to store documents for thousands of people and backing it up can cost considerable money. Having an outside service do this for you can be a money saver. Printing in mass quantities costs more than you would think. File sharing in the cloud is the way to go.

To organize Google Drive color code folders.
On a Chromebook Alt left click. On a PC, right-click. Then click on change color.

https://www.google.com/drive/

Note: If you have folders that you need to access, often put numbers in front of the name of the folder, and it will go to the top of the stack of folders.

Extension and Apps

I recommend that students old enough to have a mobile phone add apps for Google Drive, Google Calendar Google Keep and Google Classroom to their phone.

Google Drive Android
https://play.google.com/store/apps/details?id=com.google.android.apps.docs&hl=en_US&gl=US
Google Drive iPhone
https://apps.apple.com/us/app/google-drive/id507874739
Google Keep Android
https://play.google.com/store/apps/details?id=com.google.android.keep&hl=en_US&gl=US
Google Keep iPhone
https://apps.apple.com/us/app/google-keep-notes-and-lists/id1029207872
Google Calendar App Android
https://play.google.com/store/apps/details?id=com.google.android.calendar&hl=en_US&gl=US

Google Calendar App iPhone
https://apps.apple.com/us/app/google-calendar-get-organized/id909319292

Google Classroom App iPhone
https://apps.apple.com/us/app/google-classroom/id924620788
Google Classroom App Android
https://play.google.com/store/apps/details?id=com.google.android.apps.classroom&hl=en_US&gl=US

Communication

Google Classroom
This is one of the best tools to communicate and track student work.
https://classroom.google.com/ineligible
Google Classroom for Chrome:
https://chrome.google.com/webstore/detail/google-classroom/mfhehppjhmmnlfbbopchdfldgimhfhfk/RK%3D2/RS%3D76aHq9cg368hSn52btdpoDB6wKk-

Remind

This allows you to send reminders to your class.

https://chrome.google.com/webstore/detail/remind/jppddpkfhdojffabldn pdacpeoefcljp?hl=en

Seesaw: The Learning Journal

You can create a student portfolio with Seesaw.

https://chrome.google.com/webstore/detail/seesaw-the-learning-journ/adnohgfkodfphemhddnmikhflkolfjfh?hl=en

ClassDojo Extension

This is a great program to keep track of students and communicate with parents.

https://chrome.google.com/webstore/detail/classdojo-extension/mbhcppckcncdempkomncfipbddlkofio?hl=en

Memory

Quizlet

This is the best flashcard website. You can study many terms in several different ways. Quizlet has a massive database of flashcards if you do not have time to make your own on just about everything academic. The website even has an assessment component. Students can use this to test their knowledge. A stack of Quizlet flashcards can be shared with a class.

Website

https://quizlet.com/

Extension

https://chrome.google.com/webstore/detail/quizlet/bgofflgeghkhocbocio cnckocbjmomjh?hl=en

SnowLord's Quizlet Extension

This app works with Quizlet, a flashcard website that quizzes you with games.

https://chrome.google.com/webstore/detail/snowlords-quizlet-extensi/ocpkldjgfaimjjemnlppehhgdbagajhp?hl=en

Vocabulary.com
This extension helps teach you vocabulary from Vocabulary.com.
https://chrome.google.com/webstore/detail/vocabularycom/nacjcabfnee
bdomlmehblfinaepipepd?hl=en-US

Todoist: To-Do list and Task Manager
This is a very popular and highly rated to-do list.
https://chrome.google.com/webstore/detail/todoist-to-do-list-and-
ta/jldhpllghnbhlbpcmnajkpdmadaolakh

Notes

Google Keep
Create sticky notes and color code them. Reminders can be set and best
of all you can import your notes into Google Documents.
Website
https://keep.google.com/
Extensions
Google Keep Chrome Extension
https://chrome.google.com/webstore/detail/google-keep-chrome-
extens/lpcaedmchfhocbbapmcbpinfpgnhiddi?hl=en
Google Keep - Notes and Lists
https://chrome.google.com/webstore/detail/google-keep-notes-and-
lis/hmjkmjkepdijhoojdojkdfohbdgmmhki?hl=en

Mote
This outstanding extension allows you to leave voice comments in Google
Workspace. Students that have the extension can also leave voice
comments. There is a free and paid version.
https://chrome.google.com/webstore/detail/mote-voice-notes-
feedback/ajphlblkfpppdpkgokiejbjfohfohhmk?hl=en-US

Diigo Web Collector - Capture and Annotate
This Chrome extension helps you to collect information from across the
internet. You can save videos and highlighted text. This is an excellent tool
for compiling information for a report.
https://chrome.google.com/webstore/detail/diigo-web-collector-
captu/pnhplgjpclknigjpccbcnmicgcieojbh?hl=en

Google Keep Chrome Extension
This is a very popular note-taking website product made by Google. This extension will make it faster and easier to access your Google Keep account.
https://chrome.google.com/webstore/detail/google-keep-chrome-extens/lpcaedmchfhocbbapmcbpinfpgnhiddi?utm_source=chrome-ntp-icon

OneNote Web Clipper
This allows you to save a ton of information on the internet to OneNote.
https://chrome.google.com/webstore/detail/onenote-web-clipper/gojbdfnpnhogfdgjbigejoaolejmgdhk?hl=en

Sticky Notes - Just popped up!
This is a simple pop-up window of notes.
https://chrome.google.com/webstore/detail/sticky-notes-just-popped/plpdjbappofmfbgdmhoaabefbobddchk?type=ext&hl=es-419

somnote
This is a very user-friendly and free note-taking extension.
https://chrome.google.com/webstore/detail/somnote/abfmiciknijebdoghnlinglnifcobkkh?hl=en

Mic Note -Voice Recorder & Notepad
This is app will take notes and record what is being said at the same time.
https://chrome.google.com/webstore/detail/mic-note-voice-recorder-n/nhkoenoennbjnibepkjdheodiaojdgpk?utm_source=chrome-app-launcher-search

Timer
Task Timer
For elementary students, timers are an excellent way to set goals, and create structure and expectations. For special education teachers and psychologists, this is a great way to keep pace with timed testing.
https://chrome.google.com/webstore/detail/task-timer/aomfjmibjhhfdenfkpaodhnlhkolngif?hl=en-US

Chapter 3 Blended UDL

Why Blended UDL? To get an overview of UDL go here.
CAST: The UDL Guidelines: https://udlguidelines.cast.org/

Schools in large numbers are moving towards one-to-one devices. Even if a school does not use the one-on-one device model, technology is becoming the mainstream option for schools. Businesses are using word processors and spreadsheets shared online. Paper is not obsolete, but electronic documents are the standard. For schools to keep up and teach students the skills they will need to compete in a modern world, using technology is a must.

Schools are not just throwing out all paper and going electronic. Schools, in general, will have to blend the old way of teaching with the new school. The field cannot just change over immediately. The case for blended learning or combining the traditional classroom with the modern technology way of teaching is needed in education.

UDL stands for universal design for learning. The idea is that the regular education classroom should target all learners. This means teaching people who speak English as a second language, learning disabilities, anxiety disorders, economic challenges, medical disorders, and students having different learning styles. As teachers are expected to teach everyone.

Teachers that believe in and use the principles of UDL have to do several things to create opportunities for all learners. UDL teachers set an academic goal and give students various options to achieve that educational goal. Someone that is afraid of public speaking might prefer to write a paper. Someone that struggles with writing might give a speech. There are a number of options that could be used to help an individual with a vast array of challenges. Educators should present information in many different forms. In special education, we would call this multisensory teaching. Information is presented in ways that engage all senses. You see it, hear it, and then do a hands-on activity. Often you have to give real-life examples and reframe the information. An educator that can show a video, use audio, pictures, and the written form, then explain it verbally with multiple examples followed by presenting a hands-on activity. Universal design for learning is also about engagement. UDL makes the point that teachers are expected to help create student engagement. A barrier to learning is a lack of engagement or lack of interest. Technology is one of the fastest ways to increase engagement in a classroom.

UDL is also about removing barriers to learning. The whole concept of universal design for learning is giving access to the curriculum to all learners. I consider removing barriers to learning as the fourth principle of UDL. (There are only three principles of UDL)

Assistive technology, or as I like to call it, "technology" today, is part of mainstream education. The technology used with the disabled was expensive and often characterized as large and clunky. Today assistive technology is software, free, cheap, standard, on every platform, and far less stigmatizing.

Assistive technology started a number of the technologies that people use every day. Those word suggestions that appear on your phone started as word prediction for assistive technology. When you talk to your phone or tell your television controller to pick a specific channel, that technology is a direct descendant of assistive technology. Every operating system and browser has assistive technology, or today what we call "technology".

Including people with disabilities is something significant. The original law that started special education was passed in 1975. (U.S. Department of Education) Having students with various disabilities in public schools was a radical change from practices that lasted as long as everyone could remember. The inclusion movement in the 1990s that put large numbers of students with disabilities in regular education was a significant change that not everyone wanted. To this day, there are challenges when it comes to including students in regular education. With technology, those challenges get much more manageable. What I expect is that assistive technology becomes mainstream in education by being a part of blended leering.

U.S. Department of Education. (2022, March 18). *A history of the individuals with disabilities education act*. Individuals with Disabilities Education Act. Retrieved August 4, 2022, from https://sites.ed.gov/idea/IDEA-History#:~:text=On%20November%2029%2C%201975%2C%20President,and%20locality%20across%20the%20country.

Blended and Personalized Learning Vs. Blended UDL

Blended learning is characterized by combing traditional classroom teaching with online learning or computer-based learning. A teacher leads the classroom. Regular student-teacher interactions happen on a regular basis face to face. However, a significant proportion of most of the learning occurs on a computer. In some cases, students never use paper and pencil, and the teacher never uses whiteboards. The two constants are student-teacher interactions and learning are happening partly or mainly on a computer.

Personalized learning is about teaching to the specific needs of all your students. This is about individualized computer-based feedback to teach students. Each student gets individualized feedback or personal learning. This can be done with programs that give each student specific feedback tailored to each student.

Blended UDL does everything that personalized learning does and more. The general concepts of personalized learning fits well within the philosophy of blended UDL. Universal design for learning encourages engagement as a primary goal within teaching practices. Adding in high-interest learning is highly beneficial to student growth and reinforces active learning. Universal design for learning is about removing barriers. There is an overlap between to two concepts. The difference is universal design for learning gives you a blueprint to create lessons that are technology and non-technology based. Personalized learning says you are tailoring learning to each student. That means using technology. Personalized learning does not specify how you are giving the feedback. If you have prepackaged programs that give excellent feedback, then personalized learning is easy. Not every educator has such a luxury. Personalized learning does not give you the same process to follow that UDL does. The truth is you can use both in your teaching, but when creating your own lessons, UDL gives you guidance. Blended UDL is about teachers creating learning experiences that target various populations within a classroom to meet an educational goal with the use of technology.

Blended UDL is about removing barriers with technology. That means using speech-to-text, text-to-speech, word prediction, magnifiers, closed caption, and many technologies. Most of these technologies are included in your operating system and browsers. There are also free or low costs apps/extensions that work with the platform you are on. Teachers are expected to include students with any number of disabilities. Blended UDL is about giving teachers the tools to help special needs students access the curriculum. An essential principle of universal design for learning is engagement. Many challenging learners need to be actively engaged in the learning process, and just giving them technology is not enough. Blended UDL takes the principle of engagement from UDL and combines it with technology for your blended classroom.

The main point is that blended UDL takes personalized and blended learning concepts and adds to them to help various challenging populations, especially special needs students. There are a number of challenging learners that are not in special education. Many students speak English as a second language, live at or near the poverty line, and many psychological and medical disorders that impact learning go undiagnosed for decades. UDL gives teachers guidance on how to create lessons—personalized learning relies on individual feedback from technology.

Anxiety is a prevalent barrier to learning. We are only recently started to understand the impact on people's lives that have issues with anxiety. Giving students with anxiety high-interest lessons, technology to remove barriers, written directions, and allowing them to work at their own pace can help include students that quietly and often unknowingly suffer in silence.

Simply giving students specific and personalized feedback is just not enough. Providing instant feedback has tremendous benefits for students and teachers. As a profession, we have the tools and resources to accomplish so much more. We all must work with students with countless challenges in regular education. Learning the most basic concepts and tools to help challenging learners to prosper is paramount. Using such tools makes the learning process more empowering and less stressful for teachers.

But Personalized Learning and UDL are the Same Things?

Many people believe that personalized and blended learning is the same thing as blended and universal design for learning. There are similarities. Blended learning has been paired with personalized learning, but universal design for learning typically is not. So, what is the difference? The answer is the process of creating lessons is different. Universal design for learning tells you to present information in multiple ways and to have an academic goal with many paths to achieving that educational goal. It tells you to remove barriers to learning and increase engagement. If you are making your curriculum with technology to reach a diverse group of students regarding how they learn, UDL makes sense. Personalized learning tells you to tailor all the material to each student with technology but does not give you more details. With personalized learning, you give each student individual feedback with technology. However, with UDL, there are more dimensions that you can add to your teaching. Personalized learning does not consider engagement that does not involve technology or using technology to remove barriers. It does not give you the roadmap of how to blend traditional teaching with technology.

If you are going to give individualized feedback to each student all the time, you need technology. That is personalized learning. By no means should you stop using the concepts of personalized learning. I recommend using personalized learning and using UDL. There are programs that schools can buy to personalize learning. In my district, they use i-Ready and IXL, for instance. You could create a Kahoot or a Pear Deck that gives each student individualized feedback but probably not for every lesson. If you do not have the technology needed and time to create material to give specific feedback to each student, then Blended UDL is a good pathway to take. That is why you should use personalized learning and add universal design for learning to your wheelhouse. Each student can take a different path to achieve an educational goal. This often means project-oriented learning. If that is not what you do, you can create options for students in the lessons that you make. When you are starting that does not have to mean unlimited options. Giving more options to meeting a goal creates student ownership and helps students with diverse learning styles.

Applying Blended UDL

With the concept of universal design for learning, you want to create more pathways for students to reach an academic goal. To do blended learning, the student has to use technology. Many teachers have a set curriculum and believe that this is difficult to do. To completely change all your practice is difficult. To change a few lessons is not. I am not saying complete transformation. I am suggesting that you try blended UDL in increments and build on what works best. Implement what is realistic for your subject, grade, teaching style, and the students you teach.

Key Concepts for UDL

1. Removes barriers to learning (assistive technology, accommodations, written directions, work organized online)

2. Students can achieve an educational goal through multiple paths. (video, speech, writing, physical product, recording)

3. Student engagement

4. Present information in multiple forms (see, hear, touch)

Note: CAST and every other UDL author lists three principles of UDL and describes the principles with different wording. I consider removing barriers as a core concept of UDL. Feel free to read the stand explanation of UDL from CAST.

CAST: The UDL Guidelines:
https://udlguidelines.cast.org/

Math

Note: I picked math because this is a very difficult subject to implement these ideas. Having students have multiple paths to solve the same math problem and present the same problem in several different ways is problematic.

How do you create multiple pathways for success in math? Have one project where the students create and solve the math problems with real-world examples. Then present the results to the class. There might not be enough time to do this for every unit; however, you will see the power of this new model if you try this once. For a high school math teacher, this would not be realistic to do every day. However, you will see the value.

There is research to show that students that do learn math faster and perform better. The reason is they enjoy it. Positive reinforcement inspires careers. Teaching knowledge can help solve problems in the future. Inspiration creates dreams, desires, and a thirst for learning. People like control. We are hardwired this way. Do you want to have control of your classroom? As teachers, we don't like it when outsiders demand we make changes without understanding the consequences. Children and teenagers are constantly told what to do. When given the rare opportunity to do what they want, they often thrive. People remember information when it is connected to something they know or tells a story. If math is part of a story, you can see the need to remember math. Why do we remember things? Remembering is increased when connected with emotion. Some good or bad happens, and we remember. We retain information when it fits in our scheme of thinking, our world, or is part of an exciting story that we can tell. How many worksheets can you remember the problems from math class when you were in school? What were the math problems and the answers? You probably do not have an answer.

What is an example of this? Have students create geometry problems that solve real-world problems and present the class with examples. Some of the creative examples you will get will be worth remembering. The lesson will be excellent if students understand to keep it school appropriate.

What I am suggesting is that you give them more paths to learn the same information. If a teenager can take a silly meme from the internet (school appropriate) and solve a math problem that answers a question that people have, that is powerful. By giving multiple paths to learning the same information, we are benefiting from numerous psychologies that are powerful. There is research to back this up. What I am suggesting is that you give them more paths to learn the same information. There is research to back this up.

If you cannot create many paths to achieve the same academic goal, then try creating more than one for some of your lessons. The more opportunities to learn that you give to students, the better. For most subjects, this is much easier than math.

References

Bomar, M. (2009, July). Real Life Problem Solving in Eighth Grade Mathematics. Retrieved July 16, 2022, from https://digitalcommons.unl.edu/cgi/viewcontent.cgi?article=1009&context=mathmidactionresearch

Pugalee, D. (2010, March). *Writing, Mathematics, and Metacognition: Looking for Connections Through Students' Work in Mathematical Problem Solving*. ResearchGate. Retrieved August 5, 2022, from https://www.researchgate.net/publication/229480525_Writing_Mathematics_and_Metacognition_Looking_for_Connections_Through_Students'_Work_in_Mathematical_Problem_Solving

Blended UDL Templates

I am not changing universal design for learning and blended learning. I am just trying to simplify the topic with easy-to-understand concepts. A better way to say it is simplifying both and adding them together.

Think of this as a blender and add your topic for the lesson, structure, education goal, accommodations, and assessment. You can call it blended UDL because it adds UDL to technology. You can learn blended UDL by doing. Make an outline or game plan and teach a lesson with the structure that you create. Remember that you will get better with practice if you cannot fill in all the blanks the first time.

Blended learning focuses on the structure of the classroom. For many teachers, that is decided by what technology they have in their classroom. This template focuses on what teachers feel comfortable doing and can control. The best part is it is simple.

Questions with Answers

How are you going to teach the students?

Insert topic here_____

Structure

How are you going to break the time during the class into smaller structured mini-lessons?

The teacher demonstrates the topic or task.

Students work in groups or do hands-on activities.

Students complete a task on their own.

Present the information in different forms. (Multisensory teaching)

Visual

Auditory

Hands-on activity

How are the students going to meet the education goal?

How can you create more options or choices for students that are realistic for your subject and grade level?

They should have several or more options.

How is engagement being created? List examples of how you create engagement with technology and structure?

The students get instant feedback from technology.

The work is hands-on.

The students get to work in groups.

 Students play an educational game.

The students have a choice of activities.

How are you going to structure the lesson?

Several mini-lessons where the students engage in different activities.

The students have some choice of what to do.

How could you modify the lesson for student(s) that have a barrier to learning?

Students can use built-in assistive technology.

The students can look up information online or use online tools.

Students can learn from other students when working in groups.

If the students all do different projects, how do you grade them?

I have a rubric that spans several categories.

How are you going to make the directions easy to understand?

Students get a clear idea of what to do when you create good visuals and simple, straightforward directions.

Give examples of a finished product(s) that students can see or an idea that students can try. List concrete ways to crystalize understanding in pupils.

What technology are you going to use, and how will it make your lesson better?

I am going to use technology that shows the information in several different ways.

The technology used allows the students to answer questions and meet educational goals in diverse ways and provides creativity.

The technology gives students instant feedback, and as a result, students like it.

How are you going to pace the lesson?

You can change your mind later on, but you should have expectations of how long each part of your lesson will take.

What is your assessment going to be? (Formal or informational)

The evaluation could be an exit ticket, question(s), observations, task, art, graphic organizer, or outline.

Projects, quizzes, tests, and writings can also be assessments.

Blended UDL Blank Template

How are you going to teach the students?

Insert topic here_____

Structure

How are you going to break the time during the class into smaller structured mini-lessons?

How are the students going to meet the education goal?

How can you create more options or choices for students that are realistic for your subject and grade level?

How is engagement being created? List examples of how you create engagement with technology and with structure?

How are you going to structure the lesson?

How could you modify the lesson for student(s) that have a barrier to learning?

If the students all do different projects, how do you grade them?

How are you going to make the directions easy to understand?

What technology are you going to use, and how will it make your lesson better?

How are you going to pace the lesson?

What is your assessment going to be?

Chapter 4 Google for Math

One-to-one devices for math can be a real benefit. Typically, we think of one-to-one devices for subjects that do reading, writing, and research. There are many Chrome extensions for math that works well in a one-to-one device classroom.

Equation Editors

Many people do not know that Google Documents has an equation editor. This is helpful for a student that has difficulty with handwriting. It is also great for teachers that want to create crisp, clear equations that are easy to read.

Click on insert then equation.

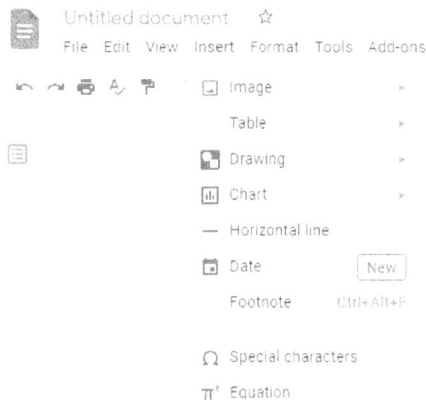

This is the equation toolbar.

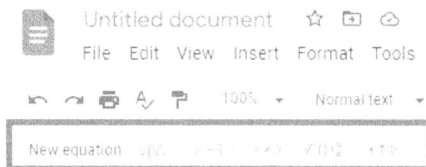

There are several equation editors add-ons. Just click on add-ons.

Add-ons   Help   Accessibility

Document add-ons   ·

Get add-ons

Manage add-ons

There are three equation add-ons for Google Documents. They are Hypatia Create, MathType, and Auto-LaTeX Equations.

If you go to the Chrome Web Store, there are math extensions that you can find. I recommend trying EquatIO from Texthelp.

Extensions For Math

EquatIO

This is a math editor that works with the Chrome browser. You can use a free version or pay for a premium version. Texthelp has made many outstanding products that can be accessed with Chrome. With so few good options for people with disabilities related to math, this app is worth consideration.

Extension

https://chrome.google.com/webstore/detail/equatio-math-made-digital/hjngolefdpdnooamgdldlkjgmdcmcjnc

Website Texthelp

https://www.texthelp.com/en-gb/company/education-blog/april-2017/math-made-digital-equatio-is-here/

Microsoft

https://www.microsoft.com/en-us/p/equatio/9p02xk38043p?activetab=pivot:overviewtab

Website EquatIO

https://m.equat.io/

Desmos Graphing Calculator

This is a graphing calculator

This is a free graphing calculator website that is free and gets outstanding reviews.

Website

https://www.desmos.com/calculator

Google Play

https://play.google.com/store/apps/details?id=com.desmos.calculator&hl=en

The App Store

https://itunes.apple.com/us/app/desmos-graphing-calculator/id653517540?mt=8

Extension

https://chrome.google.com/webstore/detail/desmos-graphing-calculato/bhdheahnajobgndecdbggfmcojekgdko?hl=en

GeoGebra

This website has many graphing and geometry features. I recommend that you look at the website to understand the many features that are offered. You can use GeoGebra on all the major platforms by downloading it or by using the website. There is a lot to like here.

Website

https://www.geogebra.org/

Extension

https://chrome.google.com/webstore/detail/geogebra-classic/bnbaboaihhkjoaolfnfoablhllahjnee?hl=en

The App Store

https://itunes.apple.com/us/app/geogebra-graphing-calculator/id1146717204

Microsoft

https://www.microsoft.com/en-us/p/geogebra-classic/9wzdncrfj48n?activetab=pivot:overviewtab

Google Play

https://play.google.com/store/apps/details?id=org.geogebra&hl=en_US

Graspable Math Sidebar

You can drag and drop equations from a webpage to this sidebar that the extension creates. You can enter equations with an equation editor and rearrange equations that you drag from a webpage. This extension even integrates with Google Classroom. I recommend looking at their website.
Website
https://graspablemath.com/learn
Extension
https://chrome.google.com/webstore/detail/graspable-math-sidebar/akhomcacccpndpgckgpkmcijkimphhmk?hl=en-US

Wikipedia with MathJax

This extension allows you to zoom into equations in Wikipedia. The app gets outstanding reviews.
https://chrome.google.com/webstore/detail/wikipedia-with-mathjax/fhomhkjcommffnlajeemenejemmegcmi?authuser=1

The Mathist - The Joy of Math

You can write your math notes better with this app.
https://chrome.google.com/webstore/detail/the-mathist-the-joy-of-ma/ehachmeohjhhbeehmhomikfanodljcnb?hl=en

IXL

IXL is a widely used online service that teaches math.
https://chrome.google.com/webstore/detail/ixl/ojpmknlmiefdmkfbfebehccibkjdihbj?hl=en-GB

Melanto Calculator Extension

This extension will add a calculator to the Chrome browser.
https://chrome.google.com/webstore/detail/melanto-calculator-extens/olhcajgllkpacioibcjiniefblpmpech?hl=en

Daum Equation Editor

This is an equation editor for Chrome.

https://chrome.google.com/webstore/detail/daum-equation-editor/dinfmiceliiomokeofbocegmacmagjhe?hl=en

Chapter 5 Google for Physical Disabilities

Deaf or Hearing Impaired

Closed Captioning

This is when words appear on the screen to read what people are saying.

Webcaptioner.com

Webcaptioner.com is a free service that turns what you are saying into live closed captioning for free. The service is web-based and can be accessed anywhere that has internet access.

https://webcaptioner.com/

Slides

Google Slides Closed Captions

With this feature, an educator can talk, and each word is stated and then displayed on your projector or computer screen. Having the visual of the presentation is excellent for reinforcement of language.

Note: PowerPoint has similar features as well if you are on Microsoft.Click to present.

Click on the three dots in the bottom right corner. Then click on captioning preferences, then toggle captions (English only) or Ctrl+Shift+C

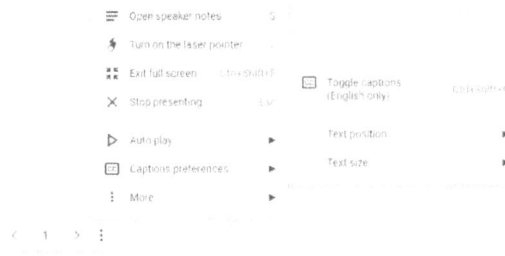

Click allow.

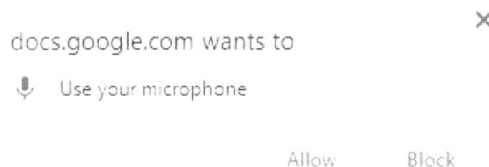

Then talk, and Slides will display in text what you are saying.

To get answers to important questions.
https://support.google.com/docs/answer/9109474?hl=en

Note: If the text is not big enough, I recommend using Google Documents voice typing. If you use an iPad as a microphone and project what you are saying on a whiteboard, it does the same thing as Google Slides Closed Captions. Except the type can be adjusted to be much larger.

Google Slides Closed Captions
https://support.google.com/docs/answer/9109474?hl=en

YouTube

YouTube has closed captioning. Creators do not have to do anything. The user must click on the "cc" on the bottom right corner of the screen to enable it.

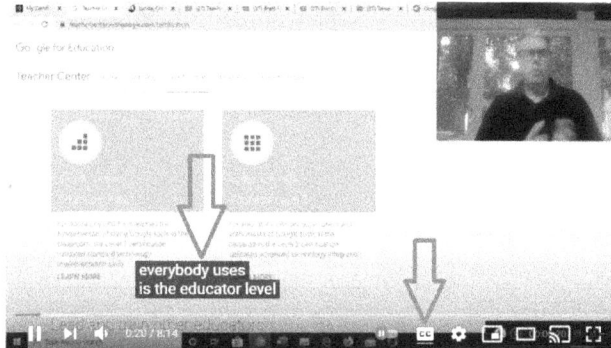

Visual Impairment

Chromebook Accessibility for the Visually Impaired
Click next to the time on the bottom right of your Chromebook.
Click on what I call the oval.

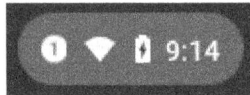

Click on the accessibility menu.

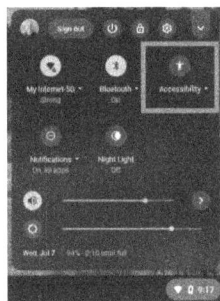

Finally, click on high contrast mode.

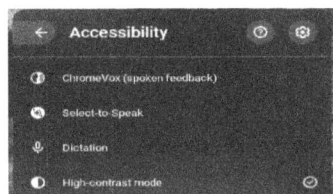

All of the colors will be inverted. However, when you talk about a screenshot of the colors.

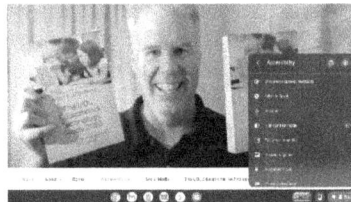

Magnifiers in the Accessibility Menu

Click on what I call the oval.

Click on the accessibility menu.

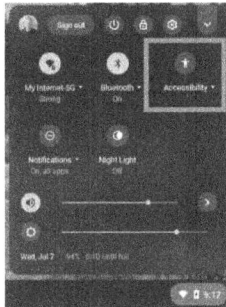

Finally, click on the full-screen magnifier.

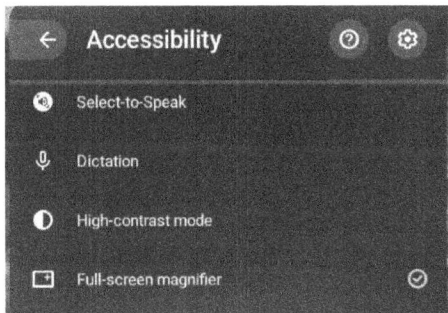

The screen zooms in on one specific area. As a result, 25% of the screen takes up 100% of the screen.

Click on what I call the oval.

Click on the accessibility menu.

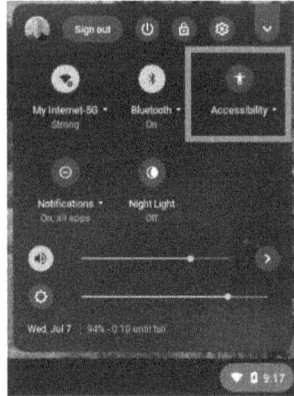

Finally, click on the docked magnifier.

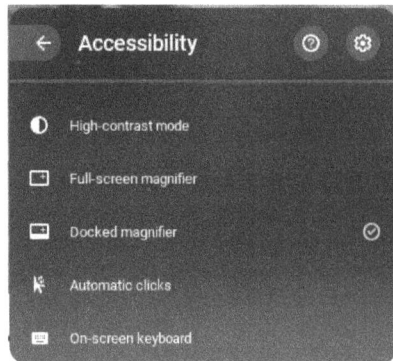

This is what it will see.

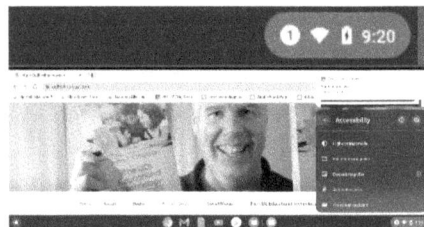

Here is another example.

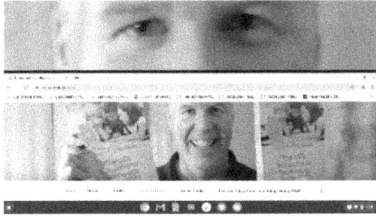

ChromeVox

This is a built-in screen reader that is designed for someone that is visually impaired. The program reads everything on the screen. That is okay if you cannot see the screen. However, if you can see the screen and want to have selected words read to you, then this is not the right option.

Click on what I call the oval.

Click on the accessibility menu.

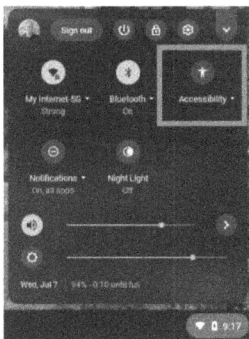

Click on ChromeVox, and you will get a blue checkmark.

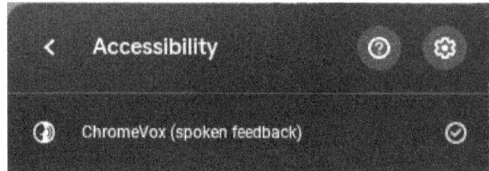

To turn off ChromeVox, repeat the same process.

Magnifiers Extensions

Zoom for Google Chrome
This app allows you to zoom into a webpage.
https://chrome.google.com/webstore/detail/zoom-for-google-chrome/lajondecmobodlejlcjllhojikagldgd?hl=en

Zoom on doubleclick
When you double-click on an item on a webpage, you will zoom in on that object.
https://chrome.google.com/webstore/detail/zoom-on-doubleclick/jkmalmidnicnnmceielaelokkdmmgkcb?hl=en

Mobility Disability

Plugin Outside Devices

The most obvious way to customize a Chromebook for a student with a mobility issue is to plug in a keyboard and/or mouse. Most keyboards and mice will work with all major platforms. To be certain that a device will work with a Chromebook, you must ask the manufacturer. When you buy the item, there are specifications that you can read that should answer this question. Worst case scenario is that you must return a device. A 30-day return policy is standard for most sellers.

Eye Gaze Mouse

When a student has a significant physical disability, an eye-gazing mouse is one of the needed devices for a computer. A person can control a mouse with their eyes if the correct software is installed. This technology used to be very expensive. However, the prices have fallen in recent years because eye gaze technology is used in gaming. Eye gaze mouse moving is still developing on Chromebooks. If you purchase a PC and use the Chrome browser, it would be feasible to use such technology. There are several companies online that you can find with a search.

Automatic Clicks

Click on what I call the oval.

Click on the accessibility menu.

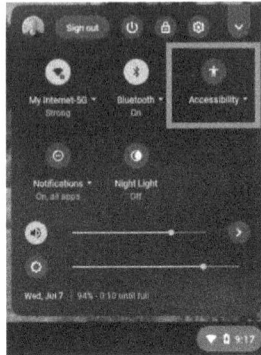

Click on the line that says automatic clicks.

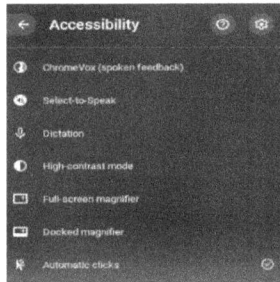

This menu will appear.

Each type of click or action can be used for a specific outcome.

On-screen Keyboard

Click on what I call the oval.

Click on the accessibility menu.

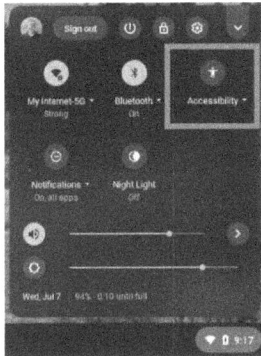

Click on the on-screen keyboard to get a blue checkmark.

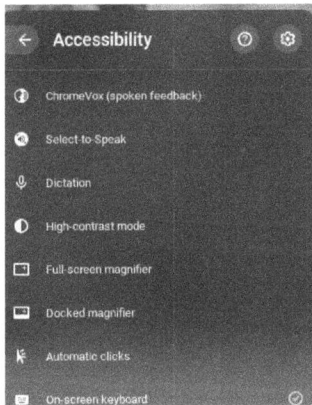

An on-screen keyboard will appear.

Google Home

Technology is going to voice. The day will soon come when you can talk to your home or a device, and everything will happen. For most people, this is cool. For people with physical disabilities, this is life-changing. The technology is developing but still effective.

https://store.google.com/us/product/google_home?hl=en-US

Chapter 6 Writing

Teaching Writing

I like to think that I am good at teaching writing because I used to hate writing, and now, I love writing. I sometimes brag about how it took me almost 18 years to pass the writing test to become a teacher. I requested the accommodation of typing the essay I had to write for the writing test, and I was refused. Granted, this was in the very early 2000 when essays were handwritten. I had been typing my papers since high school in the 1980s on a DOS computer. My issue is that my handwriting is bad. So much effort is spent on my handwriting that my overall writing suffers. In 2007 the state sent me one of my certifications without me re-taking and passing the test. In 2019 when I applied for another certification, I took the state writing test on a computer and passed. I did so without a spelling or grammar checker and no accommodations.

Test status as of September 11, 2019   Reading: Met the Qualifying Score December 8, 2001

Writing: Met the Qualifying Score August 23, 2019

Note: Ironically, 911 prompted me to take the original teacher test. The date is ominous. It took me 18 years to pass the test. An entire generation of students graduated before I was validated or vindicated by passing the test.

Note: I got my two special education certifications before the test existed.

If you want to become educated, you need to be a good writer. All colleges and graduate schools are going to ask you to write often and expect quality writing. If you want to succeed in education, you must be a great writer. Today, the good news is that you do not have to hire a lawyer to be considered a competent writer. You can use a computer without a fight.

Five Paragraph Essay

We teach a five-paragraph essay repeatedly. It is not like we do not spend enough time on it. I think most students get the underlying concepts. However, the weaker students, year after year, miss the same main points. This can be frustrating to see as an educator.

Why are we assigning writing? The goal is to teach writing. The students think the goal is to get a grade or to just do the work. Each time you write, you learn something. If you look up one thing every time you write, then you learn much more.

I simplify the five-paragraph essay so that everyone can understand it early on. Then I focus on writing.

The Five-Paragraph Essay Parts

Introduction Paragraph

What am I writing? (State in a sentence what your essay is about.)

Give background on the topic that you are writing.

or

Restate the topic sentence for main body paragraph 1

Restate the topic sentence for main body paragraph 2

Restate the topic sentence for main body paragraph 3

Thesis statement: This is your main point for the essay. This is the topic sentence for the essay or the essential question. What is the question you are asking, and how are you going to answer it?

I teach students to ask a question and then to answer the question. Paragraphs are questions that have several sentences that answer the question.

Main Body Paragraph

Topic Sentence: Ask a question and answer it. The first sentence can be a statement or a question. What question are you asking, and does the paragraph answer the question?

The three essential details that answer the question are also known as (supporting details)

1)

2)

3)

Conclusion sentence: Restate the topic sentence or restate the question as a statement.

(Repeat two more times)

Conclusion Paragraph

Restate the thesis statement, main point, or topic sentence that describes what you are writing.

Restate the topic sentence for main body paragraph 1

Restate the topic sentence for main body paragraph 2

Restate the topic sentence for main body paragraph 3

Conclusion sentence: What has the reader learned in this essay? What conclusion or main point needs to be pointed out

(Sum this up in a sentence.)

or

Restate the essential question or thesis statement as a statement.

Of course, as you become a better writer, this template is not needed. When you see students writing to write and do not think of the basic structure, it is frustrating. Not every student understands what a supporting detail means. A child can better understand a question and answer. Asking and answering a question is very simple. Restating a topic sentence, in other words, is easier to understand. Students do not always know what to write, and that is stressful to them. Every time a student writes, they have to come up with an original sentence, and that is work. If you ask questions, then answer them and restate the topic sentence three times, writing becomes simple.

Quotation

Students do not like quotations. That is ironic because it is the easiest way to improve your writing. We spend too much time complaining about plagiarism that we scare students away from using quotations. The best defense against plagiarism is creating great writers.

When you write a quotation, you are supposed to cite it. We have heard that many times. That statement is true but leaves out the most critical information.

Quotes are important statements that are so powerful that we do not want to change them. It is just more practical to present the information in its original form. Good quotes are typically short and from experts and leaders.

How do you write a quote?

1) You need to introduce the quote.

   President Kennedy said

   An expert in the field brilliantly explains this

   The best way to explain the concept of a topic sentence is

2) You need to put the statement in quotations.

3) You need to explain the quotations. If you quote Shakespeare, you need to explain the context of the quote, what the quote means, or why it is so important. There are rare exceptions. Sometimes the quote explains something so well that less of an explanation is needed.

Quotes are fantastic and powerful. Everyone should be expected to use short, powerful quotes to improve their writing.

Paraphrase

"Put this in your own words." We have all heard this quote from a teacher, and it sounds self-explanatory. However, it is not. Paraphrasing is a skill, not a catchphrase. It has to be taught and learned. It is an abstract concept that many people do wrong.

When you paraphrase, do you put everything in your own words?

Nope.

You cannot put everything in your own words. (Seriously)

You cannot change names, dates, places, or historical names of laws and battles. The words like (that, the, a, at, in, or, and us) and many other terms are not owned by anyone. Technical terms in some fields do not have appropriate synonyms. Using a synonym that is designed for a different context will make your writing worse in some cases.

If you are using some of the same words, then what does "put it in your own words mean?" You change the words that you can and the sentence structure of the writing. If a writer uses an uncommon word that you would never use, then find another word. The order of the information should be different. You can combine shorter sentences or break longer sentences into shorter sentences. The order of the information should be different and the flow of the writing. When you write about Shakespeare, you don't write in old English. No one today has a writing style that is old English. If you cite a journal article, you don't write like a PH.D. in APA style. The writing should sound and feel different. Hence, it should be in a different writing style than the piece you are taking the information. Also, make sure you cite the information that you get from a book.

Every time you write, you are supposed to learn something. It sounds like an obvious statement but do students understand this? Students often do writing to get a grade. Great writers find joy in the experience and are not afraid to look up words and ask questions to grow.

You will be no more intelligent or educated at some point in your education than the person teaching you. Someone might be an expert carpenter and not be any better at cutting wood than you. This does not mean you know more than your teacher. It does mean that you have to learn some of the information on your own for you to improve. Good learners are self-directed. For writing, what does this mean? Every time you write something, you look something up. It should be automatic, like brushing your teeth in the morning. You Google a grammar rule or question if a word is correct. This means risk-taking. Maybe you cite a blog, and you have never done that before. If you are writing and not learning, then why are you are not writing correctly.

Summaries

Summaries are when you tell a story in the shortest, most concise form. We use summaries all the time, and we do not even know it; often, when you tell a short story to answer a question, you summarize. If someone asks, where did you go? You might say that I went to the supermarket to buy roast beef for dinner. You would not say that you were stuck in traffic or stopped to get gas along the way.

When we talk, we often finish with a summary? Whenever I do public speaking, I conclude with a summary. This is just good public speaking. Good writers are concise with their words. I like to state, "say less better." The power of language is to make points, not more words. Conclusion paragraphs are a form of a summary. Summaries are paraphrasing while using a minimal number of words.

Writing Should be Fun

Everyone has a personality. I suggest you use it. This might mean telling a good story, talking about yourself or using humor. If someone has to read it, then there should be something interesting to read. Nothing is worse than reading one hundred essays about the same topic void of individualism and interest. In other words, it is TLDR or too long didn't read. Be careful when you ask a question that you do not want to hear the answer. Some students' ideas of self and fun might not mix well with school. Taking risks is not always rewarded. If you create the right structure and ground rules, then students taking risks with writing can work well.

Technology to Help with Writing

Technology should take the struggle out of writing. Everyone knows what spell checking is by now. There are many great tools to help students with writing. I see technology as part of the equation. Nothing replaces good teaching. Technology can help solve some of the issues that your students face and remove barriers.

Select to Speak

This is a free text-to-speech reader for Chromebooks that is in the accessibility settings.

Speech to Text

You have a student who has difficulty with handwriting, writing mechanics, and getting started. Someone like that would benefit from speech to text.

Dictation (Chromebook)

On Chromebooks, there is a program called Dictation in the accessibility settings.

Voice Typing (Google Documents)

Under Tools in Google Document is a free text-to-speech tool.

Apple and Windows Dictation

On Windows 10, MacIntosh, and iPads, Dictation is a free program that converts speech to text.

Taking Notes and Graphic Organizers

Google Keep is an excellent note-taking program. The best part is that you can import the notes into Google Documents.

Then simply click and drag the text.

Online Graphic Organizers

I use graphic organizers with students all the time. Sometimes I create my own. Other times I use classic graphic organizers that someone created a long time ago. One of the best ways to use a graphic organizer is by having a group of students access the same document. All the students can share their ideas by typing. Some students do better if they write their ideas instead of saying them out loud to a group. Also, when students see what others are saying, they can better formulate their thoughts. Students can learn from each other. I have seen the same idea of sharing an organized document with a group done with Jamboard and Google Slides. Both programs can be used as graphic organizers. There are even templates online for Google Slides that can be downloaded and used. Do not rule out PowerPoint templates that you find online because they can be converted to Google Slides. Upload them to Google Documents and then save them as a Google Slide.

Graphic Organizers in Word

Many free graphic organizers are made in Microsoft Word and online. You can download them and then upload them to Google Documents. Finally, convert them to Google Documents.

Graphic Organizers in Google Documents

There are free graphic organizers in Google Documents format that you can find online and make a copy.

Help with Writing Mechanics

Spelling and Grammar Checkers

LanguageTool - Grammar Style Checker
This is a comprehensive tool and the best extension in this category. I highly recommend this tool.
https://chrome.google.com/webstore/detail/languagetool-grammar-and/oldceeleldhonbafppcapldpdifcinji?hl=en

Grammarly
There is some limited spell checking with this free Chrome extension.
https://chrome.google.com/webstore/detail/grammarly-for-chrome/kbfnbcaeplbcioakkpcpgfkobkghlhen?hl=en
Grammarly App
https://app.grammarly.com/

Read & Write

This is an excellent resource if you have the paid version. The program comes with a spelling check, word prediction, and many other tools. To access the many tools, you have to use the paid version.

https://chrome.google.com/webstore/detail/readwrite-for-google-chro/inoeonmfapjbbkmdafoankkfajkcphgd?hl=en-US

Grammar Checker
You have to cut and paste from Google Documents to a window. Still, the tool is excellent.
https://chrome.google.com/webstore/detail/grammar-checker/mpeepmfabickbdbckcejbflkpfamgcon?authuser=1
https://linangdata.com/grammar-checker/

Good Documents

There is some limited spell-checking on Google Documents.

https://www.google.com/docs/about/

Citations

Google Documents

Click on tools and then citations.

Click on the citation style.

Add citations

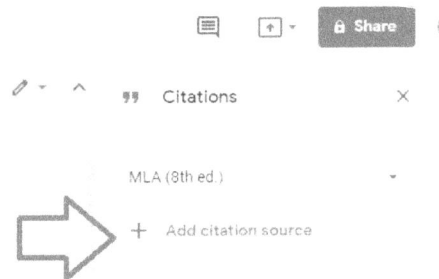

Add a source type.

Then fill out the template for reference.

NoodleTools

I use NoodleTools with my students. NoodleTools is a well-liked program that librarians and their students use to make citations. However, it is a paid program.

https://www.noodletools.com/

Citationmachine.net

When I do not have access to NoodleTools, I use Citation Machine. This is a great free online tool to create citations.

https://www.citationmachine.net/

MyBib: Free Citation Generator
This is an extension of a popular citation website. This app gets great reviews.
https://chrome.google.com/webstore/detail/mybib-free-citation-gener/phidhnmbkbkbkbknhldmpmnacgicphkf?hl=en

Chapter 7 More Tools for Google

UDL

I am a student of universal design for learning. Having been one of the first special needs students in the country back in the 1970s, I am biased toward accommodating people that don't fit in our educational system in the traditional mold. I believe in setting an educational goal and giving students multiple paths to achieve the goal. When I was in graduate school, the term was multi-sensory teaching. If I can give the information in multiple ways, it only makes sense that students can do the same to answer the question or meet the educational goal. It is not about the tool but about how you use it.

If you are giving multiple paths for students to complete an educational goal, what does that mean in the context of technology? Students can answer questions or satisfy an educational goal by writing, typing, recording sound, adding pictures, video, and with creativity when possible. Any program that gives students multiple ways to answer a question is a good tool. Any platform that allows teachers to present information in different forms is a great tool.

How to Build the Lesson

Sometimes it does not matter what platform or program you are on. It is often still the same. You are teaching the same content, and the lesson can be structured differently, but all the main parts are there. You gave a goal of what the students will learn. You start with an introduction about what the lesson is going to be. (directions) You then have the students do an activity or work. Then, you can take a class period and break it down into smaller activities that build towards the main goal of learning the material. Finally, end the lesson with some form of formal or informal assessment.

Parts

- ✓ Introduction
- ✓ Education goal
- ✓ Structured lesson
- ✓ Assessment

When we use technology, we have all the main parts or concepts of the lesson. In many cases, these ideas can translate to many online platforms to meet your academic goal, followed by an assessment.

When creating material for work on a computer, you will find you are doing many of the same things regardless of the platform. Your introduction will include directions on how to use the electronic material. The academic goal or lesson objective is the same. The type of lesson, structure, and components you create can be very similar regardless of the technology.

When you create a lesson, you have students watch a video, read an article, interpret a diagram, or perform a task. For example, if you are going to have students watch a video, answer a question together. You could put the link in a document to the video, give the question then share a copy of these documents with each group. This lesson would work as a Google Document, Slides presentation, ClassKick, Jamboard, or a Google Form. In each of these cases, the lesson would look slightly different on each program but would functionally do the same.

ClassKick

With ClassKick you can record sound, add pictures, add links to articles, links to videos, links to websites, create manipulatives, draw a picture, handwrite, type text, and use several premade features. Students can use most of these features to complete work except for the premade tools and the creation of manipulatives.

ClassKick has a free version and paid version ClassKick. The program gives you many ways to create content and for students to answer questions. This is an excellent platform for teaching special needs students and other challenging learners.

The best part of ClassKick is uploading PDFs and having the students write, type, add pictures, or record audio on the document. A teacher can create fill in the blacks with the document and even make manipulatives. The best feature of ClassKick is it allows you to upload PDFs and then quickly create dynamic online documents with the material you already have on your computer.

On the top right corner, click on the plus symbol in the green circle.

ClassKick will only show PDFs. Other types of files will not show up.

PDF Acceptable_Social_Networking
PDF Be_Comfortable
PDF Can_You_Hack_It
PDF Check_The_Privacy_Policy
PDF Citizens_Of_Cyberspace

The students can type, write, draw, record sound, and add pictures or links. Whatever the PDF worksheet used in the past, the lesson can be converted to a ClassKick.

Download PDFs from Google Slides

Download Google Documents as a PDF

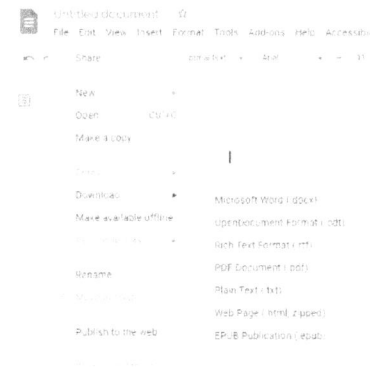

It should be noted here that you can save an MS Word document and a PowerPoint as a PDF. Then you could upload them to ClassKick.

You can also save a picture and import it as the background in ClassKick
Click on the background.

Click on import background.

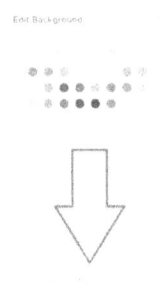

There is also a document library. Just click on browse library.

Then it will look like this.

The ClassKick Toolbar Labeled

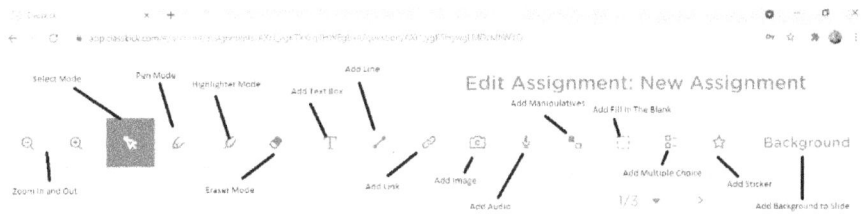

With ClassKick you can record your directions, and add screenshots, links to videos or articles. In a nutshell, you can take what you are already doing and make it dynamic and engaging. The best part is that uploading PDFs is time equitable. Meaning you get more out than what you put in.

Your students can answer a question with audio, pictures, drawing, text, highlighting, add lines and links. Students can record their answers. With the highlighting, draw a line, and pen tool, the students can annotate a document. This gives the students more options on how to answer a question. Students can add links to documents and videos on the internet.

ClassKick YouTube Channel

ClassKick has an excellent YouTube channel if you want to learn more about the product.

https://www.youtube.com/channel/UC5Yqs5875Pj9SQh0fpOyOWQ

SMAR

If you are familiar with the SAMR model, two crucial aspects of improving technology integration are finding an authentic audience and collaboration. Teachers can enhance Technology usage in ways that are not about learning the technology. It is about applying models that allow students to get feedback from others. Adding opportunities for students' collaboration and presenting to an audience provides opportunities for feedback and accountability.

Group Work

Doing group work is very different when it involves sharing documents. When most teachers think of group work, they tend to picture students looking at each other and talking. With document sharing, students can collaborate on a document. When sharing documents, each student can type on the document to help solve a problem or contribute ideas. The students can be sitting next to each other or not.

When sharing a document for group work, teachers have options. Google allows all sorts of copies to be shared. This can include Google Documents, Google Slides, Jamboards, Google Sheets, or Google Drawings. When you put the document in Google Classroom, allow all students to edit it or insert the students' names in each group. Of course, students can share out a document with each other.

When creating a dynamic document, you have options. Some Slides templates can be altered, Hyper Documents that you can create, and simple Jamboards that you can make.

Sharing in Google Classroom

To share a document, first, click on create and then assignment in Google Classroom.

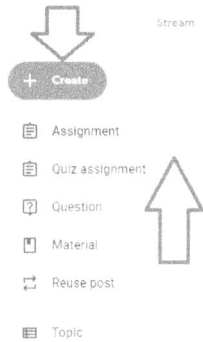

Next, in the assignment, click create and pick a document type.

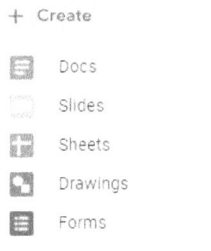

In this case, I am going to pick Slides. Sharing out documents can be done with all document types listed above and Jamboard.

Now pick **make a copy for each student**.

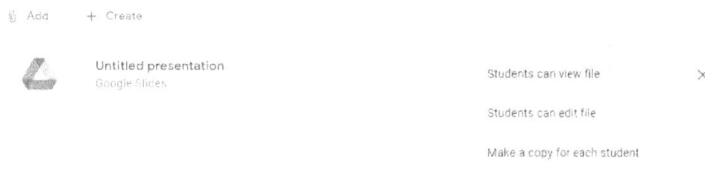

If you are doing group work, you give each student in a specific group access to one document. In that case, pick **"students can edit file."** It is a good idea to set this up ahead of time if you are breaking the class down into smaller groups.

For

All students

Points

100

Due

No due date

Most teachers in Google Schools know how to share in a post COVID world. Sharing is one way that we leverage technology. First, it creates organization for students with executive functioning issues. Sharing empowers us when we make a Google Document, Slides presentation, Jamboard, Google Form. You can share your work with them.

Sharing documents in small groups is great for student collaboration. Students today communicate via texting and writing online. By allowing multiple students to write on the same document, you allow students to model the work to each other. There are students that a not comfortable talking in a group. By adding the new dimension of being able to write allows another opportunity for students to participate.

Viewing documents looks outdated when seeing the options you have, but sometimes it is the best option. When I give directions, a rubric, or a resource document, I have students view it. If I make changes to due dates, correct a typo, or add more information, the students will see it. If you make a copy for each student, the students will see the old copy, not the new one.

Sharing Option in Google Classroom

Make a copy for each student

This is done so that each student does their own work and has a template with all the information they need in a document.

Students can view the file

Using this is when you update a document, the students see the update instead of the old document.

Students can edit the file

This is great for group work. That can be for several small groups, a student working in pairs, and everyone working on the same document.

Google Slides

Slides Templates

With Slides, content can be created in different forms. This includes pictures, videos, text, recorded sound, and any content to add a link to allow access. If a group of students has access to a Slide presentation, they can move objects you create. Slides allow you to make circles, squares, and other shapes. Teachers can make high interest games with or interactive lesson.

Slides Templates Websites

https://slidesgo.com/

https://slidesmania.com/

https://www.slidescarnival.com/category/free-templates

https://www.canva.com/presentations/templates/

Create Slides Presentations

You can create a Google Slides presentation and share it with your students. Think about what content you can share within the Slides presentation that you create and share. One of the best places to start is with the insert menu. You share a picture, text, video, diagram, chart, link, and audio files. If you share a copy for each student, then the students can insert all the exact same things.

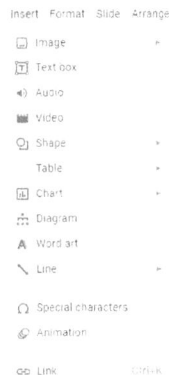

Insert Format Slide Arrange

- Image
- Text box
- Audio
- Video
- Shape
- Table
- Chart
- Diagram
- Word art
- Line
- Special characters
- Animation
- Link   Ctrl+K

Jamboard

There are some advantages to sharing out Jamboards with students. First, the interface is straightforward to use. The direction on how to use the Jamboard are less complicated compared to other types of documents. You can create multiple slides on a Jamboard if you share the Jamboard with the entire class, and this can allow each student to work on a different slide.

Sharing a Jamboard in Google Classroom

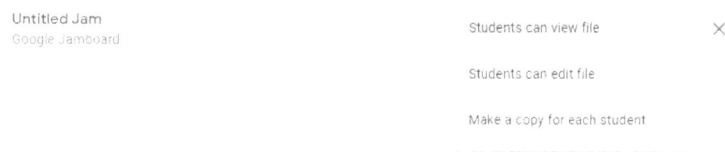

Untitled Jam
Google Jamboard

Students can view file   ✕

Students can edit file

Make a copy for each student

Reason to Share a Jamboard with the Whole Class

Each student can work on one slide or break students down into small groups to work on specific Jamboard slides.

Create a New Jamboard Slide

Just click on the arrow.

Untitled Jam

‹ 　 ›

⬆

⋮   Share

Jamboard Templates

https://www.weareteachers.com/jamboard-ideas/

https://wakelet.com/wake/IvYG3p-OAGQN2w11MM5D4

https://ditchthattextbook.com/jamboard-templates/

Hyper Documents

By inserting links and adding questions, one can create a simple hyper document. Adding links to videos, pictures, and other resources can improve the quality.

If each student makes a copy, they can add links to videos, articles, pictures, or type answers. This gives students more choice when answering questions and completing work.

When you get good at making Hyper Documents, you can take these skills to make Slides Documents you share, ClassKicks Jamboards, and other dynamic documents that your students can use.

Hyper Documents or HyperDocs

This is when many links to documents, webpages, and resources are put into a word processing document. This is typically done with Google Documents. Hence the "documents" or "Docs" in the word. This is a dynamic way to engage students in the learning process.

Websites

Several websites explain what Hyper Documents are and give you some easy to work with templates.

https://www.hyperdocs.co/start

https://hyperdocs.co/

https://sites.google.com/view/drivingdigitallearning/hyperdoc-templates

Forms

During the COVID shutdown of 2020 to 2021, some teachers used forms to give tests. Forms can be used to provide tests but have not caught with the popularity that one would think; instead of telling people to provide the tests with a form, I recommend asking a few questions. The plus of using Google Forms is that the browser on Google Chrome can be locked to prevent remote students from looking up answers. Less formal assessments can help students create awareness of what they need to learn and give feedback to the teacher about what students need to know.

Think about what you want to insert into your forms. A picture, diagram, or video can help make a Google Form more engaging. I recommend that teachers use forms to ask one to three questions to assess a classroom lesson or activity. If you start small with short quality Google Forms and build from there, you will find more success.

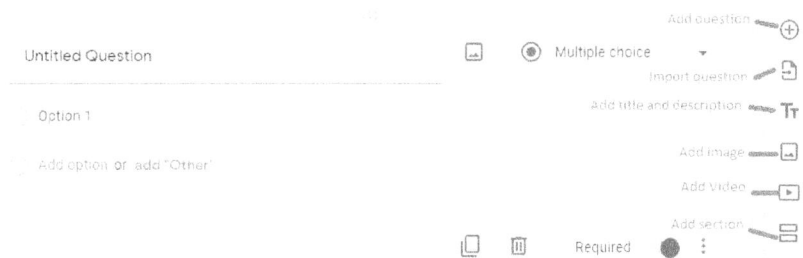

Websites

Some websites have Google Forms templates and give a basic idea of how to utilize forms best.

https://www.w3resource.com/form-template/

https://teachers.tech/google-forms-templates/

https://helpdeskgeek.com/how-to/the-10-best-google-forms-templates/

Video and Audio

Flipgrid

Flipgrid is one of the best educational technologies that is often overlooked. The best part of the platform is it is free. Teachers can ask questions, and students record their answers. Students can answer the questions as individuals or in groups. This works great for subjects that expect students to use spoken language, such as foreign language classes and speech and language therapy. With some students afraid of public speaking, Flipgrid presentations are a great alternative. Most importantly, even though it is a Microsoft product, you can log in with a Google sign-in.

Flipgrid

https://auth.flipgrid.com/signin?redirect_url=https://my.flipgrid.com/me

Flipgrid Discussion

https://info.flip.com/

Screencastify

This program could put in a separate category. Being able to record and edit video quickly is powerful. The problem is that most teachers are not interested. There was once a time video editing was problematic. It required special hardware and software. You needed a higher-end computer and special technical knowledge to precede. Those days are over. Video editing does not require special hardware, software, or memory. Editing can be done on the internet and stored in the cloud. The software is a website or an extension in many cases. The technical knowledge is more than needed for email but reasonable for an average computer user.

I use Screencastify to record directions for people all the time. The best part is that recording videos is not that time-consuming if you are not expecting perfection each time. I encourage students to create Screencasttify videos to do projects that are assigned. The problem that I face is some people do not like seeing themselves on video. Not feeling comfortable seeing yourself on video is more common than one would think. The power of being able to create videos to complete an academic goal is something that cannot be ignored. There is a free and paid version of Screencastify.

Screencastify - Screen Video Recorder
You can record the screen of your computer and even record yourself with a webcam.
https://chrome.google.com/webstore/detail/screencastify-screen-vide/mmeijimgabbpbgpdklnllpncmdofkcpn?utm_source=chrome-ntp-icon

Podcasting

One of the easiest ways to create a new dimension to your teaching is by adding Podcasting. There are many platforms and ways to Podcast. The advantage to Podcasting instead of video is that some students do not like seeing themselves on video. Students can work in groups and have a discussion. This can be a great way to have students meet an academic goal.

Best Podcast Apps and Websites for Students
https://www.commonsense.org/education/top-picks/best-podcast-apps-and-websites-for-students

Chapter 8 The Power of Instant Feedback

The first time I saw the real power of instant feedback, I changed how I teach. I was teaching computers to middle school students. A colleague of mine wanted me to test a writing program that corrected students writing. That meant I would have students write essays for a week and a half or longer instead of using a computer for other activities. Computer class is a high-interest activity, and essay writing is a lower-interest activity. I expected to see students unhappy with this change in expectations. I was surprised when they were agreeable and even happy. I saw the expression of joy in essay writing. I was shocked, confused, yet very happy.

The program that they were piloting corrected their writing and gave them feedback. The students pressed a button and were given feedback in numerous areas. They could then correct their writing. They liked writing. I could see the expressions on their faces and even enthusiasm. I have never seen children so happy to do essay writing in my life. This was all because of the power of feedback. The feedback wasn't even as fast as today with many online, blended learning programs. The power of the feedback was evident.

The traditional way of teaching writing is handwriting an essay and then handing it to the teacher. Next, the teacher makes some corrections and returns the work. In the old days, this meant lots of red ink. This might mean some comments on a Google Document in recent years. Worse than that is when a student does writing, and there is no feedback. How can you learn if you don't get any feedback?

The optimal solution is to give feedback in a way that is instant and not seen as unfavorable. When a word processor tells you to change something, it is not perceived as a huge negative to the student. If the teacher, parent, or other authority figure makes corrections, it is seen as unfavorable. Students liked the instant feedback because they got to update the assignment and create a good essay. This showed me how students love learning if they feel rewarded. The principle can work on a wide range of students and subjects.

What is Instant feedback?

Instant feedback is about making corrections or confirming correct answers immediately. People love instant feedback. The reason students enjoy this is personalization. It is the same reason people enjoy conversations where you say something and then get an immediate reaction. This power of instant feedback can make conversations go on for hours. This powerful psychological principle gets people hooked on social media, video games, and searching for information online.

Social Media is one of the best examples because the addictive powers transcend most or all demographics and ages. What social media companies do is highly effective. Data scientists compile and inspect everything we do and then figures out how to get us to spend more time on social media to sell us ads. The way to persuade us to see more ads is to convince us to return and spend as much time on the platform. The more time we spend, the more ads are seen then the rate that can be changed for those ads increases. If we argue about politics or watch a cat video, it all means interactions. Our entire economy is changing because of high-tech companies that use instant interactions and massive data to adjust to our behavior. This same power of using data and instant feedback can work for teaching with more positive results. Everyone sees the examples in our society that I am describing. I am not suggesting that we create an addiction to learning. I suggest that we can turn low-interest activities into high-interest activities with interaction, feedback, and adjustments with data. If more students learn at a high level, that is a result I can live with, and that should make us happy.

The Traditional Teaching Models

With the traditional teaching model, you test students once every two weeks. Then your students get their tests back, sometimes days later. We check to see if students did their homework, but we rely on students to correct what we write on the board. If the students are not engaged, then the effectiveness is not there. Students raise their hands only if they think they know the answer. You can call on students when they don't know the answer, but that comes with its own set of problems. If a student did not do well on a test or project, we move on to something else. The data from the test can only be used to help the students next year.

Today technology can solve these problems. Student Response Systems that are web-based can give instant feedback, correct answers, and engage all students. Microsoft Word can put red lines under incorrectly spelled words. With software and web-based services, we can give instant feedback that used to take days. If a child is not engaging, then we know it right away without embarrassment. The only problem is convincing teachers to embrace the new educational technology and letting go of the old ways.

You don't want students handwriting essays when they will be typing in the workforce in the future. That does not mean that we stop handwriting everything. The classroom should better reflect the real world. That means real-world examples and real-world technology. Instant feedback is also one of the most pleasurable ways to learn. If someone enjoys a topic, getting and keeping their attention is much easier.

The Red Ink Effect

When students write essays, they would hand them to the teacher and get them back days later with red ink. The idea was to give students feedback on spelling, grammar, sentence structure, factional information, and other writing qualities. The problem is students were often not aware that they committed many mistakes on the assignment. The corrections were coming from the teacher that they would have a personal relationship. In many cases, students spend most of the day with one teacher. That teacher's opinion meant a great deal.

They have done actual research about how red ink upsets people. Today it is almost taboo to correct students to work with red ink. Many of the students believed they were doing the work correctly. The handwritten notes from the teachers personalized the corrections. What made this worse was teachers would often circle things, and you had to look for the answers. If you misspelled a word, you go to look it up in a paper book called a dictionary. These combinations of factors made the corrections more painful than they had to be. The feedback was slow, figuring out the answers was delayed, and making the changes was painful. If you were a struggling student who did not like writing initially, this made you wish you never got out of bed.

Compare this to when you use a modern word processing program. When you use Microsoft Word and see a red, blue, or green line, it is private, not personal. We talk about personalizing teaching as being a good thing, but this is not true when being negative. If you see a mistake just between you and your computer, you are not worried about what people think. Personalizing insults makes them hurt more. When a computer instantly corrects your mistake, then it hardly feels like a mistake. Back in the day, people spell-checked with Microsoft Word after they finished writing. This was better than the red ink effect but more painful than instant feedback with instant correction. The entire point of instant feedback is to fix the mistake quickly to minimize the pain. If you were a coach and your athletes spent hours practicing something the wrong way, and you told them after they spent hours learning something wrong, how would they feel? The point is to minimize the pain and maximize the pleasure of success. Instant feedback makes correcting a happier process. Now you have confidence that it is right immediately.

What are the benefits of instant feedback?

When we track behavior in the classroom, we do so with ABA (Applied Behavior Analysis) Therapists. This includes the frequency and duration of behavior. It can also have the intensity of such behavior(s) as well. They mark down how many times a stated behavior(s) happens and for how long within a time span. With regular education students we often don't take the time to think about changing their behavior with the same methods. However, we have the technology tools, and knowhow to make changes with much less effort. The changes can start before the behaviors.

What I am suggesting is a proactive approach to gaining positive behaviors before such behaviors manifest. It might sound obvious to stop negative behaviors before starting, but we tend to think of behavior modification for students with challenging behaviors after they appear.

To understand human behavior, we have to look at the most significant modern example of behavior modification ever. That would be the rise of the internet, social media, video games, Amazon.com, smartphones, and streaming video.

What internet companies like Facebook and Google do is look at massive amounts of data. They look at all your internet habits as well as everyone else's. With the internet, there is a massive amount of data for companies to analyze. The data scientists have gotten so good at it that some people spend hours on everyday websites and engaging the platforms. We can even say that some people are addicted to social media. There are adults that order almost all their items on Amazon. Some millennials prefer to watch YouTube compared to cable television. This instant gratification has changed our culture.

There are lessons to be learned for education from the internet and marketing advancements. The first part is personalization. Facebook, Twitter, and YouTube all personalize their content to your interests. As teachers, we can't personalize the curriculum in the same way. Educational technology can personalize the content based on the student's level. Many online services do this by giving feedback and matching the students' information based on how well the students' complete tasks. With video, audio, and presentation software, students can demonstrate mastery of a curriculum or subject area through an alternative path, thus creating personalization.

Technology can also use depersonalization as an advantage. Personalization and depersonalization are opposites. If personalized learning is a big trend in education, then why would I also promote depersonalization? When we have to take negative feedback, we benefit from both instant feedback and depersonalization. A miss-spelled word in Microsoft Word is personalized because it is feedback to a specific person but depersonalized because you are being corrected by software and no humans are aware. Children look up to parental figures like teachers. When you get negative feedback from a machine where no one judges you, that is awesome. I cringe when I miss-spell a word in one of my books. I think I am not alone. If spell check makes instant changes, it is no big deal. It is like it never happened at all because it is just me and my computer.

The power of instant feedback with negative feedback is that it shortens the time the mistake exists. If you misspell a word and then immediately correct it, was it ever actually wrong? The old way of giving back a paper with red ink took longer and was much more unpleasant.

As teachers, we want to minimize the pain from negative feedback and maximize the joy derived from success. You build on success. You might learn from your mistakes, but you learn what not to do. Most people did not make a career based on what they liked least and failed most often. Success in education is what a child can do now and will use in the future.

If a student wants to do the work at home when it is not assigned, you are onto something. Being inspired by a teacher is when you want to learn and do school-related activities when there is no requirement. Most maybe all teachers have felt this joy within their subject area. If we can share this joy with students, then the odds of inspiring students are genuine.

Instant feedback reinforces correct answers in real-time. The sooner a person gets a reward or negative feedback, the better it works. This is an accepted psychological principle. The power of instant feedback is how you can create progress in the shortest amount of time. We are hardwired to react to instant feedback.

When I use student response systems like Socrative and Kahoot, I can see this first-hand how happy students are when they get instant feedback. The slight competitive nature can wake people up and engage students. It might be the only time you are sure all your students are engaged in the learning process. The hallmark of the success of instant feedback is laughter. You often see this with programs like Kahoot that are gamified and give instant feedback. My goal is to inspire as well as to educate. If you memorize large amounts of information, then you can get a good grade. However, if you enjoy the journey, then it is not like work. Positive reinforcement, success, and a feeling of meaning or progress is what we desire for students.

Instant feedback is possibly the most powerful aspect that technology and psychology can provide. Instant feedback explains the appeal of gambling as opposed to investing. In a casino, you most commonly find out if you win right away. With investments, it can take years or decades to see the outcome. Eating a cookie is instant feedback but losing weight is a severe form of delayed gratification. When a student gets face-to-face interaction, they can get instant feedback and personalization. This is impossible to do with twenty-plus students at once. With educational technology, you can give more personalized feedback. The negative feedback is private and immediate. Instant feedback is a powerful, proven psychological principle that can be effectively used with educational technology to empower students and make teachers look like rockstars.

Other Human Needs

To understand, we must re-examine our ancestral roots. When people are asked why humans are the dominant animal, they all come up with the same answers. We use tools; we have opposable thumbs and are very good at communicating. Humans are intelligent, and our ability to write information down for future generations is key to how we have evolved and learned as a society, with technology that sped up considerably. Everyone misses the key reason why humans are dominant.

Humans are the ultimate pack animal. A lion can't buy a buffalo stake on a long trip. Think about it anyone can pick up a phone when in distress and have someone come over and help you. The advantage of being a part of a large pack and having status and respect is paramount to our survival. Very few humans could live in isolation by themselves. Humans as individuals are weak. However, if you put a group on a deserted island, they will form a chain of command, rules, and people will start specialized jobs. The need to be part of a group is powerful. The desire to maintain a status to be respected is vital to getting others to help you or attract a mate. It is in our DNA.

The Three Powerful Needs

- ✓ Immediate feedback or gratification (immediate gratification)

- ✓ Need to be part of a group or pack (social need)

- ✓ The need to compete or to have higher social status by achieving.

By understanding these needs, you can understand why certain activities are reinforcing to students. Instant feedback with technology is a form of instant gratification and sometimes a way to minimize pain by correcting answers. We all have the instinct to be social members of society. This is how humans survive. The need to be social is powerful. Allowing some appropriate social interaction is beneficial. We all need to feel as if we are valuable by having social status; This can be done with achievement when we give students more opportunities to achieve, which allows for more chances to increase a feeling of social status or self-worth. Using educational technology can fulfill these important needs.

Many of these needs can be fulfilled during a lesson. We have already talked about instant feedback. Doing group work can allow students to learn from each other. This leads to the third need to achieve higher social status. When you give more opportunities for students to achieve this allows more students to gain social status and self-worth. If you had to go to a job every day and you were not good at it, how would you feel? Probably not very good. That is how lower achieving students feel. Working in a group or being social helps students to compare work, learn from each other, and complete. The status of success is meaningful. Yes, we do jobs for money. However, achieving and being fulfilled is just as important. When students have multiple pathways to success with universal design for learning, this gives them more opportunities to feel better about themselves and learn. Learning from failure can work. It just does not work as well. Failure only teaches us what not to do, and success inspires us to learn what to do and take risks in the future.

Chapter 9 Assessments

When someone says the word assessment, most people think about grades. Giving grades and assessments to get grades is a big part of teaching. However, on most assessments I give, I do not provide a grade. It sounds counterintuitive, but it is not. The traditional way we give an assessment is when we give a test at the end of a unit and give a grade. What the teachers learn from the data can help students next year in the same class.

I give assessments at the beginning of the unit. When I do a Pear Deck or a Kahoot at the beginning of the unit, I know students are paying attention. I go over the main points that I am going to teach. The data and feedback if for me. I find out what most students know and do not know early on in a lesson. The Pear Deck or Kahoot makes the assessment a game. Students start laughing, and students listen because they want to get the answer right.

For students to do well in school, it helps if they like school. For learning to be lifelong, the behavior has to be reinforced, and the learner has to see a purpose.

As a child, I was in special education. I heard things like buckle down and work harder. Making work painful only makes people think about going home. Work is not hard. Work is rewarding and beneficial. Work is what you make it. Have you ever worked with an incredible group of people? You go to work, and you love what you are doing. If you can capture a little of this, then your classroom will be the place to be. Teaching is about the journey and the destination.

Assessments are not all about grading. Assessment allows students to be self-aware of what they know and what they need to know. Once students create this self-awareness, they should be allowed to self-correct and learn what they need to know. Assessments give feedback to the teacher to affect teaching. Assessments provide opportunities for a teacher to create learning opportunities for students as well as grades.

Rubrics

One of the least popular things that I talk about at conferences is rubrics. Rubrics are important to create to have guidelines for how we grade. Rubrics make the process that we grade transparent. The problem is that most people do not follow rubrics to the letter. When you give tests, you get an exact number. With rubrics, you can measure the quality of work against a criterion. When you get good at grading with rubrics, the number you get is an accurate reflection of the work evaluated.

The first column has the criterion that you are grading. The first-row second cell through the last cell list the rating that you give each criterion. I list several examples, and you can choose what works for you. If you need a number to work with, each criterion is worth 4 points or four times four for the rating for a total of 16 points. Add up all the points and divide by 16. Many teachers use this as a guideline. Just take a set number of points from the rubric and translate that into a number from 0 to 100.

Sample Template Rubric

Criteria	Developing Below average Below average 1	Proficient Average Fair 2	Good Above average Great 3	Outstanding Exceeding Exemplar 4
Writing, style and language	The writing had many mistakes in content, structure and lacked detail	The writing has some detail with some mistakes	The writing was detailed with a few mistakes in grammar, style or content	The writing was detailed with no writing mistakes
Explained the 3 principles of UDL	Did not explain any principles of UDL	Explained the 1 principle of UDL and gave an example	Explained the 2 principles of UDL and gave examples	Explained the 3 principles of UDL and gave examples
Defined personalized learning	No definition or examples of personalized learning was given.	The definition of personalized learning was vague with no examples	Defined personalized learning but did not give examples	Defined what personalized learning is and gave examples
Define blended learning	Did not define blended learning or do so incorrectly with no examples	Gave a definition of blended learning with no examples	Gave a strong definition of blended learning with some examples	Gave a detailed definition of blended learning with many examples

Below is the rubric with standards and a paragraph explaining the lesson that I used for my ISTE Certification class.

Assessment: I want you to make a YouTube-style video (**school-appropriate**) that explains the terms: firewall, anti-virus, virus, anti-spyware, spyware, Trojan horse, phishing, keylogger, hacking, computer ethics, and password protection. It advises on how to protect your computer. The video should be about 5 minutes. You can read off an outline and hold up pictures. You can simply talk into a webcam and explain what each of these terms means and advise on how to avoid problems. If you come up with an exciting title, show good graphics, and make the presentation engaging, then points will be added for creativity. The grade will be determined by the content as well as the theme.

Note: The writing/language part of the rubric means using the proper vocabulary for the field to describe and communicate information. Students are expected to include all the key elements in the unit project to increase the possibility of getting a higher grade.

Rubric for Cybersecurity

Points	Knowledge	Content	Language	Presentation	total
5	The project shows outstanding knowledge in the content area of cybersecurity.	The content was very detailed and well thought out. It had all the essential information to demonstrate knowledge of cybersecurity.	The language was clear and detailed with strong vocabulary and expressive language that demonstrates knowledge of cybersecurity.	The presentation was outstanding and visually appealing, with many examples and detail.	
4	The project shows several good examples of knowledge of cybersecurity with one or more missing elements or lacking detail in one or more area(s)	The content was detailed but had some mistakes and/or was missing one or more essential features of cybersecurity.	The writing/language was detailed but had a few mistakes and/or lacked some appropriate vocabulary or information in one or more areas.	The presentation had several strong points with some detail.	
3	The project shows several or more missing notable elements in the content area of cybersecurity.	The content had some detail but had several mistakes and/or was missing some information and/or key points for cybersecurity.	The writing/language lacked essential terms and/or there were several vocabulary errors or was confusing and/or misleading.	The presentation had one or more missing elements and/or mistakes.	
2	The project shows little or weak knowledge of cybersecurity.	The content has several or more mistakes and/or was missing information and/or clear, consistent information.	The writing/language/ vocabulary had numerous errors with spelling, grammar, and vocabulary in several areas.	The presentation has several issues, mistakes, or is lacking in several areas.	
1	The project shows no knowledge and/or missing most or all the main points of the assignment.	The content lacked detail and/or was missing many or all main elements.	The writing/language/ the vocabulary was not done or had many mistakes in spelling, grammar and/or vocabulary.	The presentation was missing, had many errors, or did not have the most fundamental elements.	
Totals					Final Grade

Creating Assessments

Instant feedback or that small reinforcement is a form of instant gratification. Humans are pack animals and have a highly organized social order. You link that instant feedback to social interactions and the reinforcement centers will light up in your brain. When I teach, I use technology to give students instant feedback and allow them to work in small groups to learn from each other. The need for immediate gratification and the desire to be part of social order is powerful.

When it comes to doing an assessment with Kahoot, Pear Deck, or another program, you tap into instant feedback and social group activity. Everyone starts laughing. No one sees other people's answers, but they all want to compete. There is an innate desire to win.

What needs do assessment tools like ClassKick and Kahoot give us.?

- ✓ Immediate feedback or gratification (immediate gratification)

- ✓ Need to be part of a group or pack (social need)

- ✓ The need to compete or achieve. (social status)

Pear Deck

Pear Deck integrates with Google Slides. It works with Microsoft PowerPoint. In Slides, you have to add Pear Deck to start creating Pear Decks.

Click on the Pear Deck Icon. Sometimes, you might have to search for the add-on by typing the name.

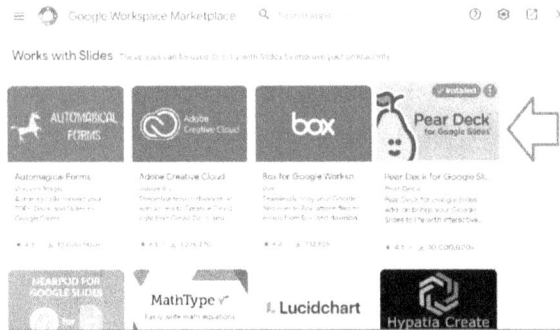

Go to the add-on, then document add-on to open the Pear Deck add-on or other add-ons, and then click on use.

The preferred way is to go to Pear Deck for Google Slides Add-on, then open Pear Deck add-on.

It should look something like this.

Click to add title

Click to add subtitle

You can make choices of many different answer types.

Text

All students can type an answer. If you use this, make sure that students sign into Pear Deck so that you can identify someone if they give a school unfriendly answer.

Just click on the lines in the gray circle. Then click on require student login.

Choice

The option choice gives you multiple-choice questions. When I use this option, I fill in the answers on the Slide. Then I enter the answers in the Pear Deck. This way, I can read the answers to the class. Then the students can see the answers again in the Pear Deck.

Number

The students answer with a number. This is great because Pear Deck graphs the answers.

Website

This is not a question type. However, you can embed a website into the Slide.

Draw

Students can draw an answer. This can be used to annotate a diagram or a picture. Less creativity is needed than one would think to use this option. The best part is that drawing is a lot of fun.

Draggable

If you have several pictures, students can drag a unique option over each picture. This is excellent for identifying what is in images.

Add Audio to Slides

Audio can be added as well. Pear Deck allows you to record, review and add audio to slides.

Teachers can look at the answers in Pear Deck to get real-time feedback. Pear Decks can be saved to look at the data later on. You could give a test on a Pear Deck. Teachers can create Pear Decks that students can do at their own pace. This allows teachers to give Pear Decks as homework.

Kahoot

Kahoot is a very popular, fun-to-use assessment program. Essentially Kahoot creates a game of questions and answers. When you do a Kahoot with a class, you can see the positive interaction. People start laughing and are engaged in the activity. Taking a test has never been so much fun.

Chrome Extensions

Pear Deck
This is a fantastic presentation program that gives feedback. This is a great way to present information and do an assessment.
https://chrome.google.com/webstore/detail/pear-deck/dnloadmamaeibnaadmfdfelflmmnbajd?utm_source=chrome-ntp-icon

Socrative Student
This is the student login for the very popular student response system.
https://chrome.google.com/webstore/detail/socrative-student/nblhpecglllndfihipmpdoikimcmgkha?hl=en

Socrative Teacher
This is the teacher login for Socrative.
https://chrome.google.com/webstore/detail/socrative-teacher/ofajgnplnindhnhnconjocimijjifaln?hl=en

Quizizz Student
This is an interactive way to quiz your class, and it integrates with Google Classroom.
https://chrome.google.com/webstore/detail/quizizz-student/iahpmdodigkpgbaolkdeelbflgeomhob?hl=en

GitHub Hovercard
This is a highly rated extension for the assessment service called GitHub. The extension gets outstanding reviews.
https://chrome.google.com/webstore/detail/github-hovercard/mmoahbbnojgkclgceahhakhnccimnplk?hl=en

Refined GitHub
This extension gets outstanding reviews. If you use GitHub, you should try this.
https://chrome.google.com/webstore/detail/refined-github/hlepfoohegkhhmjieoechaddaejaokhf?hl=en

Formative - G Suite Marketplace
This is an app for Chrome that works with the popular assessment tool.
https://workspace.google.com/marketplace/app/formative/45549193519

Edulastic
If you use this assessment tool, this extension is a consideration.
https://chrome.google.com/webstore/detail/edulastic/mmmfookngjpgda
hmnbbamplmbhleljio?hl=en

Poll Everywhere for Google Slides
You can take a poll of all your students in real-time. This is a great way to
get feedback on your students.
https://chrome.google.com/webstore/detail/poll-everywhere-for-
googl/jeehnidbmlhpkncbplipfalpjkhlokaa?utm_source=chrome-ntp-icon

Verso
This app helps you get feedback from your students.
https://chrome.google.com/webstore/detail/verso/plhfcehjadfjegljoflclmi
mcdbmnjpg?hl=en

Chapter 10 Commenting and Communication

Commenting

Good communication skills are vital when working with special needs students. Giving concise, easy-to-follow directions is paramount. You can comment with your students via comments in Google Documents, Google Slides, and Google Classroom. Being able to keep records of repeated communication is essential. As teachers, we have to contact parents in time equitable ways as well.

When we are talking about leveraging technology, that means using it more effectively. Commenting back and forth with your students or having students that work in a group that comments back and forth with each other are fantastic examples of leveraging technology. Comments can be added to Google Documents, Slides, Sheets, and Google Drawings. Having students and educators on writing can help correct mistakes and can allow students to ask questions. Google Classroom allows for communication to happen in numerous areas of the program.

Commenting on Documents

One of the best and most overlooked functions of using Google Workspace is comments. You highlight something and comment. You can also type a + or a @ then choose someone's email. That will alert the person of the comment. The person will know that the comment is for them.

Commenting does several things. First, it documents your communication with the student. Short, simple notes can bring attention to significant changes or deadlines. Many students need tasks broken down into smaller parts. When students comment back and forth, it documents each student's contribution to group work. Group work on documents can help students model work and classroom norms to each other.

Highlight the word, sentence, paragraph, or phrase. Then click on the square with a plus symbol on it.

Type the @ or + to direct a comment to a specific person.

Note: Slides, Sheets, and Drawing also allows commenting.

Commenting on Google Slides

You can share comments in Google Slides in the same way that you do in Google Documents. With some creativity, you can create some interesting lessons. Start by adding pictures, diagrams, embed videos, adding questions, and having students answer their induvial copies. Consider using online templates as well.

Click on the box with the plus symbol, then type the comment.

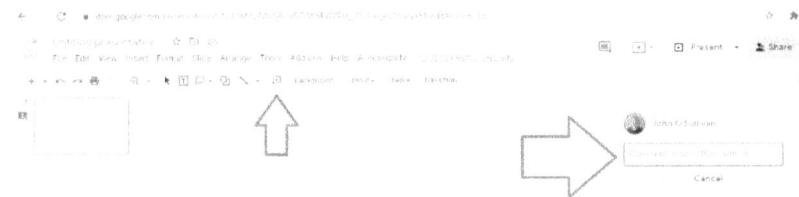

Commenting on Google Drawing

Very few people think of using Google Drawings. However, you can share a drawing with each student or with a small group of students. This is great for having students label a diagram. Answering a question by drawing a picture is engaging. I incorporated art in my teaching practice and have had success doing so. When students are stressed, having them start by drawing pictures calms them down and creates focus. It sounds insane until you try it, but it works.

First, click on the object you want to make a comment on. Then click on the box with the plus sign and, finally, comment.

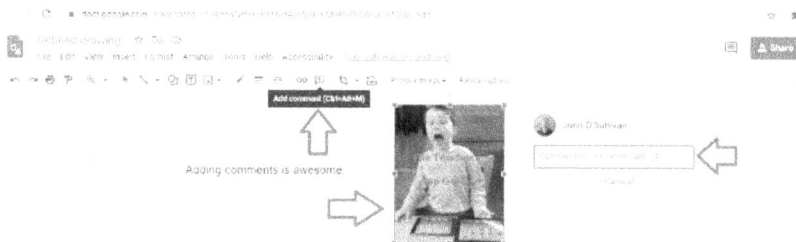

Teachers can comment on students' work. However, if you have students work in groups, they can comment to other students to help facilitate corrections.

Teaching Empathy

Talking about empathy in a computer book seems totally out of place. However, this is the most important lesson that we can teach the young. Under the ISTE citizen, standard 3a talks about empathetic behavior. As special educators, we think about social skills often. Doing the work in school is only part of the equation. People have to grow up and learn how to get along with others.

I teach students how to fix computers, and I spend a week teaching empathy. Sounds crazy? The biggest problem when you are fixing a computer is dealing with people that are stressed out. If you have good listening skills and get good information, this can help you fix the computer and sidestep some stressful situations. As teachers, we often have to teach or model social skills. However, if you have a lesson on social skills, people tend to get insulted, including individuals that lack social skills.

Empathy Map

There are several versions of empathy maps out there. The map asks the same basic questions. What does the person say? What does the person do? What is the person thinking? How does the person feel?

Says	Thinks
Does	Feels

I start teaching empathy with self-empathy. I have students fill out the map for themselves based on the last time their computer broke. Yes, we need to be self-empathic. Then after they fill it out, I ask what they could have done differently? If the answer is nothing, that is okay. At least we are thinking about it. Lack of self-awareness is a big problem. Being self-aware and empathetic are essential skills.

Google Classroom

Telling teachers to make comments in Google Classroom sounds obvious and time-consuming. As teachers, we must correct people and break larger tasks into smaller ones. We need to be gentle when we have to say something that might come across as unfavorable. Most humans are sensitive. This includes adults. Communication that is not face to face can sometimes cause miscommunication. Even with all the issues, written communication is practical. The reason is that you can address minor problems. People tend not to remember small mistakes if they are corrected quickly. When concerns grow, and authority figures confront us, that creates stress. Giving students feedback is an essential task for teachers.

Google Guardian

Google Classroom creates reports for a parent based on what they request. Parents get notifications of their student's progress. One of the top complaints of parents is they did not know about the problem. Contacting parents is time-consuming. Once parents are added to Guardian, the information gets sent out without the teacher doing any extra work. I would recommend talking to your administration and the technology department before using Google Guardian.

Google Classroom App

Taking five minutes to have secondary students load the Google Classroom on their phones will pay dividends. Educators see mobile phones as being distractions. Students should not be using phones during class. However, if they have their phones with them, it is good to add the Google Classroom app. The hope is they will check their phone for their homework and due dates. Adding the app to their phone makes their schoolwork very accessible. If that means they remember one due date, it is worth it.

Chapter 11 Google Add-ons

Introduction to Add-ons

Add-ons are one of the hidden gems within Google Workspace. Everyone should know where to find add-ons. Most teachers never use add-ons or even look at them. There is a great deal of value in add-ons on various programs in Google Workspace. Teachers should consult with the technology department before using any of these add-ons. In some cases, they may be blocked. It is also a good idea to consult technology professionals to make sure that the add-ons are safe and approved by your district.

Google Slides

Click on add-ons, then document add-ons.

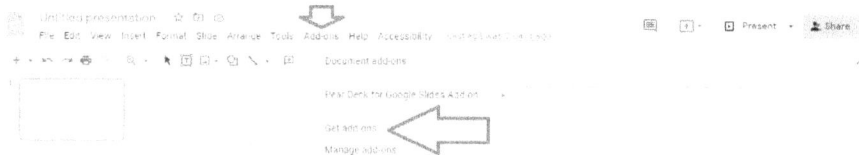

There are many more add-ons, and new ones are being added all the time. Please take the time to look for yourself.

Notable Slides Add-ons

MathType
Create math equations in Google Slides or Google Documents with this amazing add-on. Math teachers who share a Google Document or Google Slides can use this add-on with their students.

Pear Deck
Google Slides presentations can be turned into interactive assessments. Pear Deck has a group mode to assess the class's knowledge on a topic. Teachers have set Pear Deck to an induvial mode for homework or do at your pace actives.

Hypatia Create
This is an equation editor that works at a fast speed. This add-on can be used with Google Documents, Slides, and Forms.

Easy Accents – Slides
Foreign language teachers can add accents to Slides presentations.

Grackle Slides
This add-on checks for the accessibility of a Slides presentation.

Bjorn's Quiz Decks Studio
Create quizzes with this add-on. This add-on gets excellent reviews.

Slides QR Codes

You can create QR codes for your Slides presentations. This is great to make links more accessible.

Google Documents

To access Google Document add-ons, click on add-ons, then get add-ons.

Here are some notable add-ons. Please take the time to check if other add-ons will help you and your students. Also, please remember to check with your information technology department before installing add-ons.

Math

Hypatia Create
This is an equation editor that works at a fast speed. This add-on can be used with Google Documents, Slides, and Forms.

MathType
Create math equations in Google Slides or Google Documents with this excellent add-on. Math teachers who share a Google Document or Google Slides can use this add-on with their students.
Auto-LaTeX Equations
Turn math equations into images to be used in your documents.

Easy Accents – Slides
Foreign language teachers can add accents to Slides presentations.
Automagical Forms
Create multiple choice questions for your Google Documents.

Doc To Form
Add text from your Google Document into a Google Form.

Google Sheets

Click on add-ons, then get add-ons.

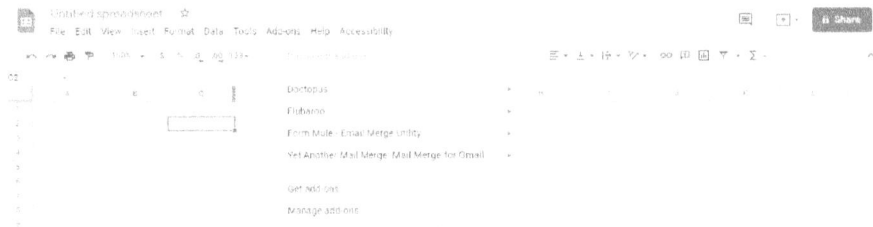

Autocrat
Create mail merges for emails. If you have a spreadsheet of contacts, you can create a mail merge to send a similar email to your contacts.

Flubaroo
This add-on helps you grade assignments.

Doctopus
With this add-on, you can mass copy and share documents from Google Drive with a group of students.

CoRubrics
Rubric creating tool.

Presence for Meet
This is an excellent way to keep track of who attended your Google Meets.

www.ingramcontent.com/pod-product-compliance
Lightning Source LLC
Chambersburg PA
CBHW081238020426
42331CB00013B/3222